FISH AND SEAFOOD

Mark Ainsworth

IDENTIFICATION · FABRICATION · UTILIZATION

Join us on the web at
culinary.delmar.com

FISH AND SEAFOOD

Mark Ainsworth

IDENTIFICATION · FABRICATION · UTILIZATION

Australia • Brazil • Japan • Korea • Mexico • Singapore • Spain • United Kingdom • United States

The Kitchen Professional, Fish and Seafood: Identification, Fabrication and Utilization
Mark Ainsworth

The Culinary Institute of America

President: Dr. Tim Ryan '77

Vice-President, Continuing Education: Mark Erickson '77

Director of Publishing: Nathalie Fischer

Editorial Project Manager: Margaret Wheeler '00

Editorial Assistant: Shelly Malgee '08

Photography: Keith Ferris, Photographer
Ben Fink, Photographer

Vice President, Career and Professional Editorial: Dave Garza

Director of Learning Solutions: Sandy Clark

Acquisitions Editor: James Gish

Managing Editor: Larry Main

Product Manager: Nicole Calisi

Editorial Assistant: Sarah Timm

Vice President, Career and Professional Marketing: Jennifer McAvey

Marketing Director: Wendy Mapstone

Marketing Manager: Kristin McNary

Marketing Coordinator: Scott Chrysler

Production Director: Wendy Troeger

Production Manager: Stacy Masucci

Senior Content Project Manager: Glenn Castle

Art Director: Bethany Casey

Technology Project Manager: Chirstopher Catalina

Production Technology Analyst: Thomas Stover

© 2009 Delmar, Cengage Learning

ALL RIGHTS RESERVED. No part of this work covered by the copyright herein may be reproduced, transmitted, stored, or used in any form or by any means graphic, electronic, or mechanical, including but not limited to photocopying, recording, scanning, digitizing, taping, Web distribution, information networks, or information storage and retrieval systems, except as permitted under Section 107 or 108 of the 1976 United States Copyright Act, without the prior written permission of the publisher.

> For product information and technology assistance, contact us at
> **Professional & Career Group Customer Support, 1-800-648-7450**
> For permission to use material from this text or product, submit all requests online at **cengage.com/permissions**.
> Further permissions questions can be e-mailed to **permissionrequest@cengage.com**.

Library of Congress Control Number: 2008931927

ISBN-13: 978-1-4354-0036-8

ISBN-10: 1-4354-0036-4

Delmar
Executive Woods
5 Maxwell Drive
Clifton Park, NY 12065
USA

Cengage Learning is a leading provider of customized learning solutions with office locations around the globe, including Singapore, the United Kingdom, Australia, Mexico, Brazil, and Japan. Locate your local office at **www.cengage.com/global**

Cengage Learning products are represented in Canada by Nelson Education, Ltd.

To learn more about Delmar, visit **www.cengage.com/delmar**

Purchase any of our products at your local bookstore or at our preferred online store **www.cengagebrain.com**

Notice to the Reader
Publisher does not warrant or guarantee any of the products described herein or perform any independent analysis in connection with any of the product information contained herein. Publisher does not assume, and expressly disclaims, any obligation to obtain and include information other than that provided to it by the manufacturer. The reader is expressly warned to consider and adopt all safety precautions that might be indicated by the activities described herein and to avoid all potential hazards. By following the instructions contained herein, the reader willingly assumes all risks in connection with such instructions. The publisher makes no representations or warranties of any kind, including but not limited to, the warranties of fitness for particular purpose or merchantability, nor are any such representations implied with respect to the material set forth herein, and the publisher takes no responsibility with respect to such material. The publisher shall not be liable for any special, consequential, or exemplary damages resulting, in whole or part, from the readers' use of, or reliance upon, this material.

Printed in Canada
4 5 6 7 17 16 15 14

Contents

RECIPE CONTENTS   x

ABOUT THE CIA   xii

AUTHOR BIOGRAPHY   xv

ACKNOWLEDGEMENTS   xvi

INTRODUCTION   1

1 FISHING METHODS   2
Fishing Methods   3

2 FIN FISH: QUALITY CHARACTERISTICS, STORAGE, AND HANDLING   10
Purchasing Fresh Fish   11
Storage   13
Market Forms for Fin Fish   14

3 SHELLFISH: QUALITY CHARACTERISTICS, STORAGE, AND HANDLING   18
Shellfish Purchasing, Storage, and Market Forms   19

4 FIN FISH IDENTIFICATION   32
Anchovy   33
Arctic Char (*Salvelinus alpinus*)   35
Barramundi (*Lates calcarifer*)   36
Bass, Chilean Sea (*Dissostichus eleginoides*)   37
Bass, Hybrid (*Morone chrysops-m. saxatilis*)   38
Bass, European Sea (*Dicentrarchus labrax*)   40
Bass, Black Sea (*Centropristis striata*)   41

Bass, Striped (*Morone saxatilis*)   43
Blackfish, Tautog (*Tautoga onitis*)   44
Bluefish (*Pomatomus saltatrix*)   45
Bream (*Sparus auratus*)   46
Catfish (*Ictalurus punctatus*)   47
Cod (Family: Gadidae)   49
Dogfish (*Squalus acanthias*)   56
Eel (*Anguilla rostrata*)   58
Flounder (Order: Pleuronectiformes)   59
Grouper, Red (*Epinephelus morio*)   61
Halibut   62
Herring (*Clupea harengus*)   63
John Dory (*Zenopsis ocellata*) (*Zeus faber*)   65
Lingcod (*Ophiodon elongatus*)   66
Mackerel   67
Mahi Mahi: (*Coryphaena hippurus*)   69
Monkfish: (*Lophius americanus*)   70
Mullet, Red (*Mugil cephalus*)   72
Opah (*Lampris guttatus*)   73
Orange Roughy (*Hoplostethus atlanticus*)   74
Ocean Perch, Atlantic (*Sebastes marinus*)   76
Pompano (*Trachinotus carolinus*)   77
Porgy (*Pagrus pagrus*)   79
Red Drum (*Sciaenops ocellatus*)   80
Sablefish (*Anoplopoma fimbria*)   81
Salmon, Atlantic (*Salmo salar*)   82
Salmon, Coho (*Oncorhynchus kisutch*)   84
Salmon, Chum (*Oncorhynchus keta*)   85
Salmon, Pink (*Oncorhynchus gorbuscha*)   86
Salmon, Sockeye (*Oncorhynchus nerka*) or Copper River Salmon   86
Sardines (*Sardinella aurita*) (*Sardina pilchardus*) (*Harengula jaguana*)   88
Skate (*Raja batis*) (*Raja binoculata*) (*Gymnura micrura*)   89
Smelt, Rainbow (*Osmerus mordax*)   91
Snapper (Genus: *Lutjanidae*)   92
Sole (Family: Achiridae)   95
Sturgeon   97
Swordfish (*Xiphias gladius*)   98
Tilapia (*Tilapia nilotica*)   100
Tilefish (*Lopholatilus chamaeleonticeps*)   101
Trout, Rainbow (*Salmo gairdneri*)   103
Tuna Overview   104
Turbot   111
Wolffish (*Anarhichas lupus*)   112

5 SHELLFISH IDENTIFICATION   114

Classifications of Marine Animals   115
Abalone (Family: Haliotidae)   117
Clams   118
Conch, Queen (*Strombus gigas*)   125
Crab   126
Cuttlefish (*Sepia officinalis*)   132
Langostino (*Cervimunida johni, Munida gregaria, Pleuroncodes monodon, Nephrops norvegius*)   134
Lobster   134
Mussel, Blue (*Mytilus edulis*)   138
Mussels, Green Lip (*Perna canaliculus*)   139
Scallops   139
Sea Urchin (*Strongylocentrotus fransiscanus, Strongylocentrotus droebachiensis*)   142
Shrimp   144
Shrimp Varieties   145
Surimi   149

6 CEPHALOPOD AND OTHERS IDENTIFICATION   150

Octopus (*Octopus dofleini, Octopus vulgaris*)   151
Oysters   152
Oyster Varieties   153
Squid (*Loligo illecebrosus, Loligo opalescens, Loligo pealei*)   156

7 FIN FISH FABRICATION   158

Fabrication   159

8 SHELLFISH FABRICATION AND TOOLS OF THE TRADE   172

Shellfish Fabrication   173
Tools of the Trade   186
Knife Sharpening   188

9 AQUACULTURE   192

Fish Farming Techniques and Methods   194

10 SANITATION: SAFETY AND SANITATION, STORAGE AND HANDLING 198

How Fresh Is Fresh Fish? 200
Time and Temperature 201
Storage 204
Purchasing 205
Food Safety and HACCP 205
Food-Borne Disease 207

11 CURING, BRINING, SMOKING, RAW, AND CAVIAR 216

Salt: Sodium Chloride (NaCl) 218
Curing 221
Smoke 228
Caviar 234
Raw 239

12 NUTRITION AND UNDERSTANDING COOKING METHODS AND INGREDIENTS 250

Nutrition 251
Understanding Cooking Methods and Ingredients 254
Keep it Simple 258

13 RECIPES 272

Ceviche 275
Salt Cod Fritters 276
Conch Fritters 277
Crab Cakes with Creole Honey-Mustard Sauce 279
Creole Honey-Mustard Sauce 281
Garlic Shrimp 282
Clam Sauce 283
Egg Pasta 284
Shrimp Tempura 285
Tempura Dipping Sauce 286
Sushi Rice 287
Japanese Hand Vinegar 288
Wasabi 288
Nigiri Sushi 289
Japanese Omelet for Nigiri Sushi (Tamago) 290

Nori-Roll Sushi (Maki Sushi)   291
Miso Soup   293
Dashi   294
Shrimp Bisque   295
New England Clam Chowder   296
Mussels Marinière   297
Bouillabaisse   299
Rouille (Garlic and Saffron Mayonnaise)   301
Salade Niçoise   303
Trout with Sautèed Mushrooms   305
Dover Sole Meunière   307
Shrimp with Tomatoes, Feta, and Oregano   309
Paella   311
Base Recipe for Shallow Poached Fish   315
Fillet of Flounder with White Wine Sauce   317
Grilled Salmon with Ginger Glaze   319
Pan-Fried Cod   321
Fish and Chips   323
French Fried Potatoes   325
Lobster Thermidor   326
Cold-Smoked Salmon   327
Gravlox (Dill Cured Salmon)   328
Salmon Rillette   329
Fish Stock   330
Mousseline Forcemeat   331

GLOSSARY   332

BIBLIOGRAPHY   337

INDEX   338

PHOTO CREDITS   349

CIA CONVERSION CHARTS   350

Recipe Contents

Ceviche **275**

Salt Cod Fritters **276**

Conch Fritters **277**

Crab Cakes with Creole Honey Mustard Sauce **279**

Creole Honey-Mustard Sauce **281**

Garlic Shrimp **282**

Clam Sauce **283**

Egg Pasta **284**

Shrimp Tempura **285**

Tempura Dipping Sauce **286**

Sushi Rice **287**

Japanese Hand Vinegar **288**

Wasabi **288**

Nigiri Sushi **289**

Japanese Omelet for Nigiri Sushi **290**

Nori-Roll Sushi **291**

Miso Soup **293**

Dashi **294**

Shrimp Bisque **295**

New England Clam Chowder **296**

Mussels Marinière **297**

Bouillabaisse **299**

Rouille **301**

Salad Niçoise **303**

Trout with Sautèed Mushroom **305**

Dover Sole Meunière **307**

Shrimp with Tomatoes, Feta, and Oregano **309**

Paella **311**

Base Recipe for Shallow Poach Fish **315**

Fillet of Flounder with White Wine Sauce **317**

Grilled Salmon with Ginger Glaze **319**

Pan-Fried Cod **321**

Fish and Chips **323**

French Fried Potatoes **325**

Lobster Thermidor **326**

Cold-Smoked Salmon **327**

Gravlox **328**

Salmon Rillete **329**

Fish Stock **330**

Mousseline Forcemeat **331**

ABOUT THE CIA

THE WORLD'S PREMIER CULINARY COLLEGE

The Culinary Institute of America (CIA) is the recognized leader in culinary education for undergraduate students, foodservice and hospitality professionals, and food enthusiasts. The college awards bachelor's and associate degrees, as well as certificates and continuing education units, and is accredited by the prestigious Middle States Commission on Higher Education.

Founded in 1946 in downtown New Haven, CT to provide culinary training for World War II veterans, the college moved to its present location in Hyde Park, NY in 1972. In 1995, the CIA added a branch campus in the heart of California's Napa Valley—The Culinary Institute of America at Greystone. The CIA continued to grow, and in 2008, established a second branch campus, this time in San Antonio, TX. That same year, the CIA at Astor Center opened in New York City.

From its humble beginnings more than 60 years ago with just 50 students, the CIA today enrolls more than 2,700 students in its degree programs, approximately 3,000 in its programs for foodservice and hospitality industry professionals, and more than 4,500 in its courses for food enthusiasts.

LEADING THE WAY

Throughout its history, The Culinary Institute of America has played a pivotal role in shaping the future of foodservice and hospitality. This is due in large part to the caliber of people who make up the CIA community—its faculty, staff, students, and alumni—and their passion for the culinary arts and dedication to the advancement of the profession.

Headed by the visionary leadership of President Tim Ryan '77, the CIA education team has at its core the largest concentration of American Culinary Federation-Certified Master Chefs (including Dr. Ryan) of any college. The Culinary Institute of America faculty, more than 130 members strong, brings a vast breadth and depth of foodservice industry experience and insight to the CIA kitchens, classrooms, and research facilities. They've worked in some of the world's finest establishments, earned industry awards and professional certifications, and emerged victorious from countless international culinary competitions. And they continue to make their mark on the industry, through the students they teach, books they author, and leadership initiatives they champion.

The influence of the CIA in the food world can also be attributed to the efforts and achievements of our more than 37,000 successful alumni. Our graduates are leaders in virtually every segment of the industry and bring the professionalism and commitment to excellence they learned at the CIA to bear in everything they do.

UNPARALLELED EDUCATION

DEGREE PROGRAMS

The CIA's bachelor's and associate degree programs in culinary arts and baking and pastry arts feature more than 1,300 hours of hands-on learning in the college's kitchens, bakeshops, and student-staffed restaurants along with an 18-week externship at one of more than 1,200 top restaurant, hotel, and resort locations around the world. The bachelor's degree programs also include a broad range of liberal arts and business management courses to prepare students for future leadership positions.

CERTIFICATE PROGRAMS

The college's certificate programs in culinary arts and baking and pastry arts are designed both for students interested in an entry-level position in the food world and those already working in the foodservice industry who want to advance their careers. A third offering, the Accelerated Culinary Arts Certificate Program (ACAP), provides graduates of baccalaureate programs in hospitality management, food science, nutrition, and closely related fields with a solid foundation in the culinary arts and the career advancement opportunities that go along with it.

PROFESSIONAL DEVELOPMENT PROGRAMS AND INDUSTRY SERVICES

The CIA offers food and wine professionals a variety of programs to help them keep their skills sharp and stay abreast of industry trends. Courses in cooking, baking, pastry, wine, and management are complemented by stimulating conferences and seminars, online culinary R&D courses, and multimedia training materials. Industry professionals can also deepen their knowledge and earn valuable ProChef® and professional wine certification credentials at several levels of proficiency.

The college's Industry Solutions Group, headed by a seasoned team of Certified Master Chefs, offers foodservice businesses a rich menu of custom programs and consulting services in areas such as R&D, flavor exploration, menu development, and health and wellness.

FOOD ENTHUSIAST PROGRAMS

Food enthusiasts can get a taste of the CIA educational experience during the college's popular Boot Camp intensives in Hyde Park, as well as demonstration and hands-on courses at the new CIA at Astor Center in New York City. At the Greystone campus, CIA Sophisticated Palate™ programs feature hands-on instruction and exclusive, behind-the-scenes excursions to Napa Valley growers and purveyors.

CIA LOCATIONS

MAIN CAMPUS—HYDE PARK, NY

Bachelor's and associate degree programs, professional development programs, food enthusiast programs

The CIA's main campus in New York's scenic Hudson River Valley offers everything an aspiring or professional culinarian could want. Students benefit from truly exceptional facilities that include 41 professionally equipped kitchens and bakeshops; five award-winning, student-staffed restaurants; culinary demonstration theaters; a dedicated wine lecture hall; a center for the study of Italian food and wine; a nutrition center; a 79,000-volume library; and a storeroom filled to brimming with the finest ingredients, including many sourced from the bounty of the Hudson Valley.

THE CIA AT GREYSTONE—ST. HELENA, CA

Associate degree program, professional development programs, certificate programs, food enthusiast programs

Rich with legendary vineyards and renowned restaurants, California's Napa Valley offers students a truly inspiring culinary learning environment. At the center of it all is the CIA at Greystone—a campus like no other, with dedicated centers for flavor development, professional wine studies, and menu research and development; a 15,000-square-foot teaching kitchen space; demonstration theaters; and the Ivy Award-winning Wine Spectator Greystone Restaurant.

THE CIA, SAN ANTONIO—SAN ANTONIO, TX

Certificate program, professional development programs

A new education and research initiative for the college, the CIA, San Antonio is located on the site of the former Pearl Brewery and features a newly renovated 5,500-square-foot facility equipped with a state-of-the-art teaching kitchen. Plans for the 22-acre site include transforming it into an urban village complete with restaurants, shops, art galleries, an open-air *mercado,* an events facility, and expanded CIA facilities, including a demonstration theater and skills kitchen.

THE CIA AT ASTOR CENTER—NEW YORK, NY

Professional development programs, food enthusiast programs

The CIA's newest educational venue is located in the NoHo section of New York's Greenwich Village, convenient for foodservice professionals and foodies alike. At The Culinary Institute of America at Astor Center, students enjoy courses on some of the most popular and important topics in food and wine today, in brand-new facilities that include a 36-seat, state-of-the-art demonstration theater; a professional teaching kitchen for 16 students; and a multipurpose event space.

AUTHOR BIOGRAPHY

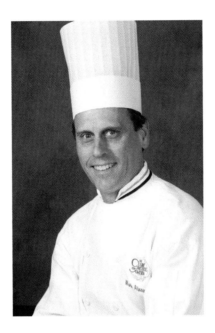

Mark Ainsworth is a professor in Culinary Arts in the Continuing Education Department at The Culinary Institute of America (CIA). A 1986 graduate of the CIA, he holds dual certification from the CIA and the American Culinary Federation as a ProChef Level II (PC^{II}) and Certified Chef de Cuisine (C.C.C.). Chef Ainsworth is a Certified Hospitality Educator (C.H.E.) who has also taught courses in the college's degree programs.

Prior to returning to his alma mater as a member of the faculty, Chef Ainsworth was executive chef at Pussers Landing in Tortola, British Virgin Islands and for Clipper Cruise Lines in St. Louis, MO. He was also chef-de-partie at the renowned Le Bernardin in New York City, and held chef positions at The Grill Room at the Hotel Bayerischer Hof in Munich, Germany; the Charleston Marriott Hotel in Charleston, WV; and the Presbyterian Center in Holmes, NY.

Chef Ainsworth was a member of the CIA faculty team which won the coveted Marc Sarrazin Cup at both the 1996 and 1997 Salon of Culinary Arts in New York. He has also earned a silver medal at the International Dietetic Cooking Competition in Bad Wörishofen, Germany and an honorable mention in the Nestlé Chocolate Competition in White Plains, NY.

In addition to his degree from the CIA, Mark Ainsworth has earned a Bachelor of Arts in Media-Communications from the University of South Carolina in Columbia, SC.

ACKNOWLEDGEMENTS

In 1966 when I was a young boy, my favorite pastime was fishing with my father on the lakes of New York Hudson Valley. With two sisters and a brother, there was always competition for his attention, but once a year we each enjoyed our own special time with him. My choice was always to travel from our home along the south shore of Long Island for a weekend of camping and fishing. Always an early riser, I was up and out of our cabin at the crack of dawn to fish the lake for perch, sunfish, bass, and pickerel. Although I had a love for fishing, I never mustered the courage to remove the fish from the hook. So with fishline in tow, I would sprint up to the cabin and wake my father from his quiet country sleep by dangling the fish in front of his face. This was a time before catch-and-release so we kept most of the fish. The bass, sunfish, and perch would be filleted and fried up in bacon drippings over the wood fire and eaten for breakfast. Boney pickerel were headed, gutted, and packed in a cooler of ice and saved until we got home, where they were destined for the blender and made into fish cakes. This book is dedicated to my father, who grew up along Lake Champlain and often shared precious memories with me of happy times spent with his father in old wooden boats on the pristine lakes of Vermont.

The administration, staff, students, and faculty of The Culinary Institute of America have all contributed to this book through their enthusiasm for culinary education and love of food. And special thanks go out to Chef Corky Clark and Brad Matthews for all their assistance.

There are others who have inspired me along the way: My longtime friend and fishing companion, Fred Frost, whose love of both fishing and books was the initial catalyst for this endeavor. To Darby, Kaleigh, and Patrice who put up with my absence and nurtured me through the entire project. To my mother, whose persistent attention through life gave me the skills I needed to get the job done. Thanks to my colleagues' Chefs Crispo, Deshetler, Von Bargen, and Kamen, as well as Tama Murphy, and Mark Erickson in the Continuing Education department of The Culinary Institute of America, for sharing their knowledge and encouragement. The great photographs throughout the book are crafted by two of the best, Ben Fink and Keith Ferris; they have been a pleasure to work with. Keeping it all organized has not been easy, and Shannon Eagan and Maggie Wheeler have both worked long and hard multitasking with great passion and precision.

Lastly, they say that every good writer has a great editor and I could not have written this book without the assistance of Rose Occhialino, a lover of books, who in her 87 years has probably never read and reread so much about fish. Her constant enthusiasm, focus, and attention to detail can be found swimming through every page like a school of invisible needle fish.

INTRODUCTION

Seafood is flavorful, nutritious, and readily available worldwide. Oceans cover over 70 percent of the earth and their abundant bounty has always been appreciated and utilized as a sustainable, renewable resource. In the past several hundred years as the world's civilizations have become more technologically advanced, larger amounts of seafood have been taken, often to the detriment of the individual species. Fishing techniques and processing and storage advancements have outpaced the sea's ability to reproduce, and world governments have been left to regulate and control what little is left. Although land-based protein sources have always made up the bulk of our diet, as they become more expensive and questionable from a safety and health standpoint, we will need to revert to the sea for salvation.

With seafood consumption rising yearly, it is important to understand the complexities of the products we purchase and serve. No one can deny our right to fish, but it should be done correctly, using modern regulated techniques that are sustainable. Knowledge and education regarding seasonality, aquaculture, nutrition, safety, and sanitation are essential. Chefs, restaurateurs, and home cooks all need a better awareness of the complexities involved in our consumption of the sea's bounty.

This book encompasses myriad features of fish: its history, seasonality, its nutritional importance in our diet, as well as health concerns such as toxins and allergies associated with its consumption.

Also explored are the various fishing techniques employed, identification of the important species that inhabit our waters, procedures for preserving freshness from harvesting to market, and suggestions for the best cooking methods for each type of fish.

Understanding all of these factors of the seafood industry is of great importance: It will not only increase the bottom line, it will enhance the customers' enjoyment, raise awareness of the fragility of the ecosystem, and contribute to the protection of a precious natural resource.

FISHING METHODS

FISHING METHODS

Since humans have inhabited the earth, the sea has supplied a nutritionally satisfying, high-protein, low-fat food. Because of these riches, people settled along the coasts, lakes, and rivers of the world. Many fishing techniques have been employed throughout time, and certain basic methods largely have gone unchanged. Others have been refined and adapted to the modern age, but fishing's basic concept is fairly straightforward: to remove as many fish as needed with the least possible effort. Before preservation methods, most of the catch had to be consumed immediately or it would spoil. With the advent of refrigeration, and the motorization of fishing fleets, the race was on to develop more efficient ways of harvesting seafood.

Nature works in harmonious ways and seems to supply the tools for all species to survive. Fishing gear has advanced from hook and line to modern factory ships capable of processing fish nonstop 24/7. The danger in these high-tech methods is that we will take more than we require and disrupt the delicate balance of nature. Regulations and sustainable methods must be developed and enforced to ensure long-range management of the world's oceans for future generations.

From the beginning, hooks and spearheads were crafted out of different materials including bone, shell, and stone. Fibrous plant material was pulled and spun into suitable line capable of hauling a fighting fish from the water. One line became two and soon large nets were constructed to gather the fish that escaped the lines. Netting materials have advanced from natural to synthetic and, with the use of winches, are able to haul large quantities of fish very rapidly.

A variety of fishing methods are used to catch both fin fish and shellfish. Some are designed to target specific species and others are designed to increase yield. Understanding how the fish we eat is harvested and insisting on sustainable methods will go a long way to protecting the ocean from unnecessary damage and either limit or utilize unwanted bycatch.

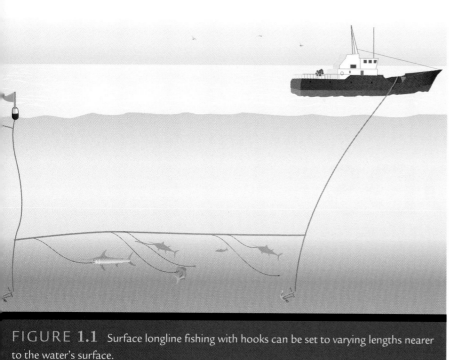

FIGURE 1.1 Surface longline fishing with hooks can be set to varying lengths nearer to the water's surface.

LONGLINE FISHING

Longlining is one of the most productive methods of catching fish. Lines of varying lengths, some as long as 50 miles, are rigged with baited hooks at set intervals throughout the water. Bottom fish such as cod, halibut, and monk fish are caught with anchored lines set horizontally and are marked with surface buoys for tuna and mahi mahi, whereas swordfish lines are set closer to the surface (see Fig 1.1). Horizontal lines are also employed and anchored to the bottom and buoyed on top.

Longlining is a controversial fishing method because it indiscriminately catches unwanted fish species as well as marine mammals and birds, in particular the albatross. Methods deemed friendly to sea birds include fishing at night and setting streamers on the lines to scare the birds away. Eliminating, minimizing, or utilizing waste from fish fabrication is another step in the right direction. New methods are turning the fish by-products into usable fish meal on board the vessel. This encouraging development goes a long way toward true sustainability.

GILLNETTING

Gill nets are long walls of nets set close to or below the surface (see Fig 1.2), on the bottom, or at various depths depending on species and location. They can be easily located along a known migration path to catch large quantities of fish. Varying in mesh size, these nets are invisible to the fish as they swim into them. Once their heads and gills go through the net, they become entangled and die, which drastically affects the quality of the fish, so speed in harvesting is essential.

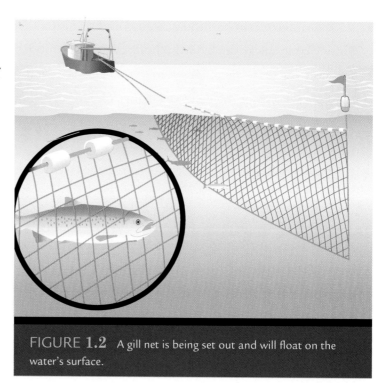

FIGURE 1.2 A gill net is being set out and will float on the water's surface.

CHAPTER 1 · FISHING METHODS

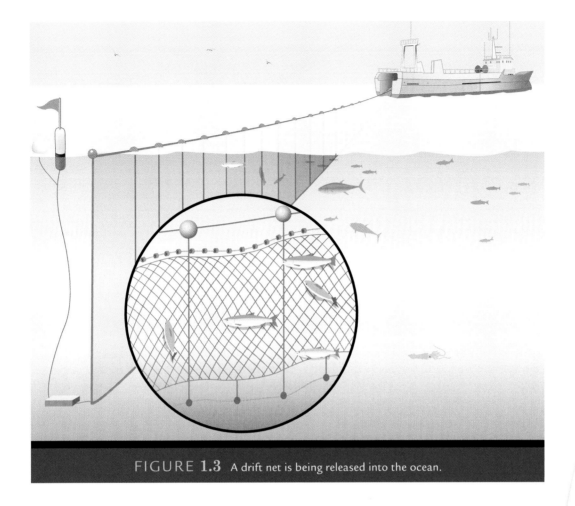

FIGURE 1.3 A drift net is being released into the ocean.

DRIFT NETS

Trapping fish in the same way as gill nets, drift nets are not affixed to anything and silently move with the tide, entangling the fish (see Fig. 1.3). Used at sea to catch squid, tuna, salmon, and other valuable species, these nets have prompted the United Nations to recommend a global moratorium on large-scale high-seas drift netting to protect the large pods of dolphins and turtles from becoming entangled in nets up to 3,000 yards long.

Easily lost, and invisible, they are referred to as ghost nets; they drift and fill up with fish until the weight causes them to sink to the bottom of the sea. Once the entangled fish are consumed by other marine life, the net floats back up to the surface repeating the process. Unfortunately, modern nylon nets do not disintegrate but stay intact, rising and falling in the sea. A disadvantage of both drift and gill nets is the indiscriminate catch of species.

TRAWLING

Trawling is a method of fishing that pulls different sized nets through the water to capture various species of fish and shellfish. Boats can operate in tandem, pulling large nets through the water, or a single vessel can use a beam, which holds the net open as it is

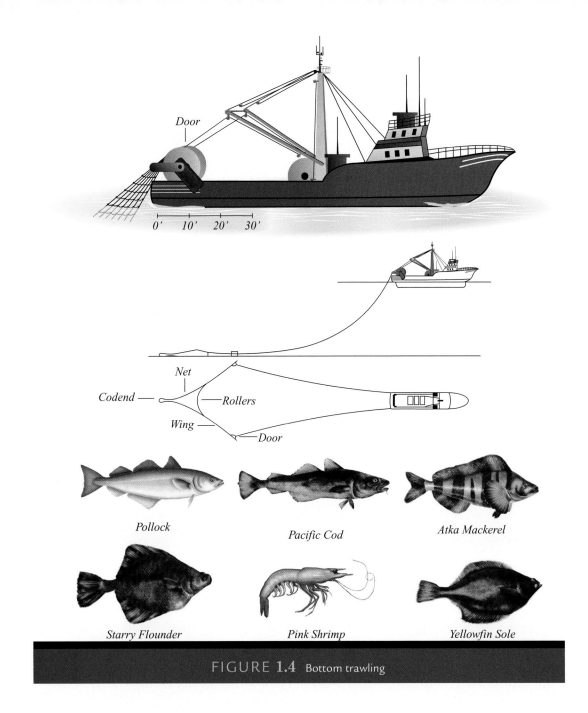

FIGURE 1.4 Bottom trawling

dragged along the ocean bottom or at various depths (see Fig 1.4). Bottom trawlers have chains attached that stir up the seabed and force ground fish up into the waiting net. Trawling nets are controversial because of the damage they cause to the ocean floor.

MIDWATER TRAWLING

Midwater trawling deploys a large cone-shaped net from the stern of the boat and pulls it through the water scooping up anything in its path (see Fig 1.5). Once full, the net is hauled onboard and the fish are placed in the hold. Unwanted bycatch and damage to the fish as they are lifted onto the vessel are disadvantages of this method.

TROLLING

Trolling utilizes lures or baited lines from the stern of the boat to capture valuable game fish such as tuna, mahi mahi, and sailfish. Weights are connected to wire lines

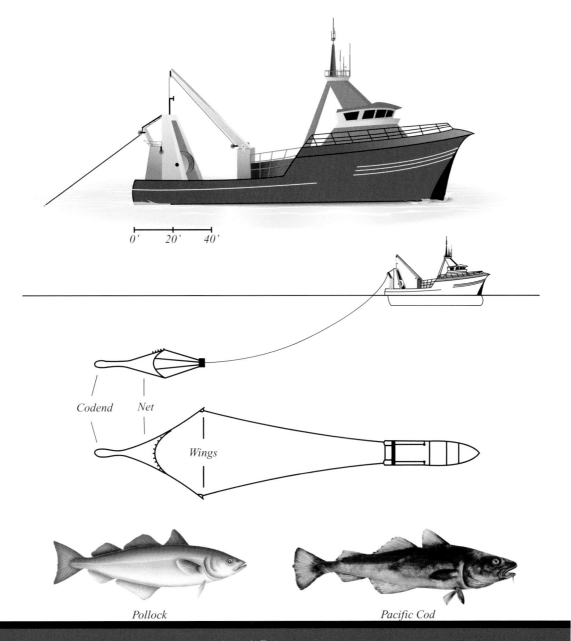

FIGURE 1.5 Midwater trawling

with 15 to 20 leaders, each of which is pulled behind the boat (see Fig 1.6). Each line can also be rigged individually and winched in to quickly recover the fish alive. This method is especially beneficial for tuna because their body temperature can increase drastically during the fight and proper bleeding and immediate cooling are important to the value of the fish. Fish can also be more easily targeted by utilizing specific jigs and live bait.

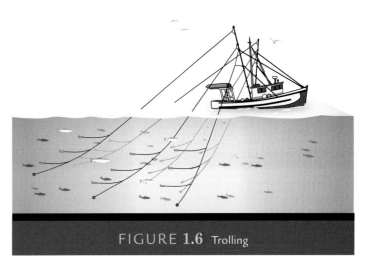

FIGURE 1.6 Trolling

FISHING METHODS

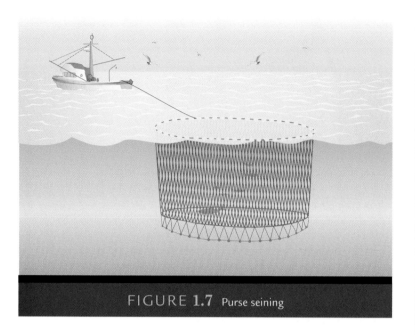

FIGURE 1.7 Purse seining

PURSE SEINING

Purse seining encircles schools of fish with a wall of net that is then pursed (drawn together) on the bottom, trapping the fish (see Fig 1.7). The entire net is brought to the side of the vessel and the fish are pumped or scooped onboard. Targeting large shoals of tuna and mackerel, this method became controversial in the 1970s when dolphins were deliberately encircled to facilitate catching the tuna with which they congregated.

FISH TRAPS, OR POTS

Lobster, crab, and fish are caught using various sized pots made of wire, metal, wood, and line. The pot is baited, thrown overboard, and sits on the bottom attached to a buoy (see Figs 1.8a through 1.8c). The entrance is designed to prevent escape from the trap. An advantage of this method is that it is highly selective; everything is caught alive with little or no bycatch or habitat destruction. Pot sizes vary; with Alaskan red crab, pots are able to hold hundreds of pounds of crab. Fish traps are especially popular throughout the warm calm waters of the world for their ability to catch specific varieties of fish.

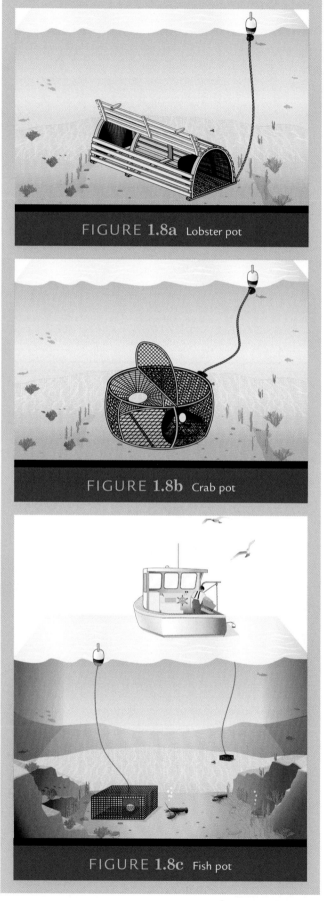

FIGURE 1.8a Lobster pot

FIGURE 1.8b Crab pot

FIGURE 1.8c Fish pot

DREDGING

Primarily used for shellfish such as clams, scallops, mussels, and oysters, a dredge is a metal basket with a type of rake or teeth assembly that aids in removing mollusks from the seabed (see Fig 1.9). Clam dredges at sea are very large and must be towed from a sizable vessel. Modern dredges pump pressurized water in front of the rakes to loosen the silt and churn up the shellfish. Towed from bars off each side of the boat, the number of baskets or dredges deployed from a single vessel may reach several dozen depending on the catch. Dredging is controversial because it can tear up and disrupt the sea bottom, as well as have a negative effect on the natural sediment of the spawning habitat of shellfish.

FIGURE 1.9 Dredging the ocean floor for clams or scallops.

DIVERS

Divers utilizing scuba gear, or air pumped from the surface, collect a wide range of shellfish from the sea bottom (see Fig 1.10). Scallops collected this way are referred to as day boats because the divers harvest and return on the same day. Due to the high cost of harvesting, these items command a premium market price.

FIGURE 1.10 Divers in scuba gear collecting scallops.

FIN FISH: QUALITY CHARACTERISTICS, STORAGE, AND HANDLING

PURCHASING FRESH FISH

Because fish are extremely perishable, it is important to understand the quality characteristics of freshness.

QUALITY CHARACTERISTICS (see Figures 2.1a through 2.1d)

- Clear eyes
- Red gills
- Fresh aroma
- Firm flesh
- Moist and shiny fillets with tight flake
- No belly burn

Purchasing whole fish rather than fillets or steaks allows for a longer shelf life and makes for easier quality assurance. The first prerequisite is correct storage. Attention to detail throughout the "flow-of-goods chain" is the mark of a reputable and knowledgeable seller and typically ensures a consistent product. All of the senses are used when choosing fish, but smell is vital in determining quality. Fish should smell like the mist of an ocean wave, clean and briny.

FIGURE 2.1a Gills should be maroon to bright red.

FIGURE 2.1b Round fish should be in the position they assume when swimming, with their cavities full of ice.

FIGURE 2.1c Scales should be intact and the flesh should be firm, not soft and spongy.

FIGURE 2.1d Cavity smells clean and briny, has a bright color, and is free from discolored blood. If the guts were not removed immediately, belly burn will occur and discoloration will be evident.

A fishy, stale, or iodine odor is a clear indication that the fish is old or has been improperly stored. Aroma is a prime indicator of freshness for fillets or steaks, which because of their processing, lack many of the other indicators. The skin should feel moist and look shiny, not sticky, dry, or show ice burn. Certain species such as skate, catfish, and squid are naturally slimy, which is not a detriment to quality. Scales, if there are any, should be firmly attached. Fins and tail should be moist, intact, and full, suggesting proper handling. The flesh should be firm, not mushy and soft; when pressed with your finger, there should be no visible or lingering indentation. Once filleted, the muscle structure should appear intact and firm without a soft or spongy texture, a factor that can also be determined by how the knife cuts, not pulls, through the fillet. The eyes are the windows to freshness; they should be moist, clear, and slightly bulging. As the fish ages, the eyes dry out and sink back into the head. Bright red to maroon gills, without any sign of gray or browning, indicate a quality product. The absence of gills is a sure sign that the fish is not fresh; avoid purchasing them. When purchasing drawn or pan-dressed fish, inspect the cavity; it should be moist, bright, and free of belly burn, which occurs when evisceration is delayed, causing stomach enzymes to break down the flesh.

Fish should arrive the way they were ordered. If they are whole fish, their entire body should be immersed in ice in a swimming position, including the cavity. Round fish placed on their sides during shipment can be easily damaged. Smell is one of the best indicators of freshness.

INDICATORS OF POOR QUALITY FISH

- Fishy, stale, or "off" aroma
- Mushy flesh, open-flaked fillets
- Dry-looking skin or fillet
- Discoloration
- Dark gills
- Belly burn

STORAGE

Fish should be received packed in crushed or flaked ice (Fig. 2.2). After checking for freshness, immediately place all seafood under refrigeration. The best way to store round fish is to pack their cavities with ice, place the fish in the swimming position, and surround the entire body with more ice (see Figs. 2.3a and 2.3b). Proper drainage

FIGURE 2.2 Whole fish should arrive packed in crushed or flaked ice.

FIGURE **2.3a** Round fish should be packed in crushed or flaked ice; ice cubes of other shapes will bruise the fish.

FIGURE **2.3b** Round fish should be in the position they assume when swimming, with their cavities full of ice and the sides in contact with ice, not other fish.

FIGURE **2.3c** Flat fish, such as flounder and halibut, should be stored in crushed ice at angles. Flat fish can be easily damaged by stacking them in bins or tubs.

is very important as is the need for more ice as necessary, since melting will occur. Flat fish are best stored with their thicker, dark side down, at a slight angle, and packed in ice (see Fig. 2.3c). This is critical to preserving freshness and maintaining quality.

MARKET FORMS FOR FIN FISH

The seafood market is highly competitive and to generate sales, suppliers and processors are continually seeking new products and ways to repackage and update existing lines. Fish sticks are now fish fingers; fish nuggets and even fish French fries have become popular. The United States has seen a steady rise in seafood consumption in the last decade, with an average of 16.5 pounds consumed per person in 2006. Approximately $20.5 billion is spent yearly on seafood, leading the United Nations to project a 40-million-ton global shortage by the year 2030. With this increased consumption comes competition and the influx of a variety of choices on the retail and commercial level.

MARKET FORMS

WHOLE FISH

Also referred to as round fish, this is the entire fish as it comes out of the water, its head, scales, and viscera intact. From a cost perspective, fish in the round are usually the most economical because they have not been processed. This may seem to be cost effective, but scaling, gutting, fabrication, and waste disposal takes space, labor, and money, which increases the cost. Also, individual fish have drastically different meat-to-bone ratios, and you can actually pay more for the whole fish than you would for fillets.

From a utilization standpoint, whole fish allows for maximum product utilization; head and bones can be made into fish stock and turned into profitable soups and stews (see Fig 2.4). Quality of whole fish can be easily measured by looking at the eyes, gills, and especially at the belly for signs of bloating, which occurs when the viscera are not removed promptly. High levels of bacteria in the stomach and other internal organs quickly deteriorate the fish, especially with improper storage and age. This is particularly problematic with species such as salmon, mackerel, and tuna, so these fish are normally gutted soon after landing.

Round fish such as salmon and red snapper have two distinct fillets (see Fig 2.5). Flat fish such as halibut and flounder can be fabricated by removing either one large fillet per side or two separate fillets per side. Larger flat fish such as halibut can be further broken down into smaller fillets called *fletches* and generally have a very high meat-to-bone ratio. Commercial fillets are sold fresh in plastic containers, and on the retail level they should be displayed on ice with a moisture-resistant material separating the fish from contact with ice.

Fillets offer convenience but are the most expensive form of fresh fish available. Fillets are typically sold boneless, but some pin bones may remain, necessitating further processing. Portion control is an advantage of fillets, allowing the operator to order specific sizes, ensuring consistency in plating and food cost.

FIGURE 2.4 Fish purchased head-off and gutted may or may not be scaled but should still have their tails and fins attached, which can be later used to produce a fish stock. Headed and gutted fish also allow for more flexibility in fabricating steak cuts by cutting through both fillets and across the backbone. Shipping and in-house costs will be less than fish in the round, but if bones are not used for stock, then calculating your as-purchased (AP) and edible portion (EP) cost is recommended.

FIGURE 2.5 Fillets are the entire side of the fish removed from the backbone from head to tail and are the most popular form available. Skin should be intact for many species such as red snapper and sea bass to identify that the fish received is the one that was ordered.

Freshness indicators for fillets

One disadvantage of purchasing fillets is that many of the freshness indicators have been obliterated.

Careful inspection using the following guidelines will ensure a quality product.

- Fillets should be moist, shiny, and glistening.
- Shapes should be consistent, well trimmed, and free of most bones.

- Excess water in the container indicates freezing and thawing.
- Aroma should be fresh, clean, and briny, not fishy.
- Firm, stiff, and dry fillets indicate that they have been previously frozen.
- Packaging should be intact and well sealed.
- Color should be consistent without traces of blood, yellow, or gray.
- Skinned fillets should be totally free of skin.

Portion size
Portion-size fillets come fabricated in specific sizes and weight and are available more readily frozen than fresh.

Skin on or off
As mentioned, it is advisable to purchase certain high-cost species with the skin on to avoid any product substitution. Common fish like flounder, monkfish, and cod are normally skinned immediately after filleting; whereas others with softer flesh are left skin-on to avoid the fillets breaking up during processing.

Steaks
Fish steaks are cross-section cuts, available bone in or bone out. Steaks may also contain the belly flap and pin bones, especially when cut from round fish such as salmon. This "round" steak, including the belly flap, can be secured into a cylindrical shape with cotton twine for easier cooking and presentation. Depending on the fish and cooking method, skin can be left on or removed after cooking. Larger flat fish such as halibut can also be cut into various shaped steaks. These are not fabricated into steaks in the same manner as round fish but are filleted and cut into squares or rectangles.

Cross-cut steak is an economical way to cut a round fish. It involves less labor and offers premium product utilization. They are also easily portioned and can be sold by the piece in retail operations. When cutting a steak from a round fish it is not advisable to utilize the tail section, which tapers down into a very thin portion with a thick bone. Use this tail meat for other items such as fish stews, mousseline, or rillettes.

A disadvantage to cross-cut steaks is that the bones are still intact and may be undesirable to customers or guests, although bones do add flavor when cooking.

Steak preparation
Determine the appropriate portion size and the number of cuts necessary for the specific fish. Cut through the top fillet to the backbone with a large knife. Increasing the knife pressure, cut through the backbone and bottom fillet until the steak is separated from the fish. Larger fish may require the use of a rubber or wooden mallet to penetrate the backbone, but be careful not to damage and bruise the bottom fillet.

SHELLFISH: QUALITY CHARACTERISTICS, STORAGE, AND HANDLING

SHELLFISH PURCHASING, STORAGE, AND MARKET FORMS

Shellfish can be purchased in a variety of forms to meet the needs of customers and operators. In addition to fresh, frozen, and canned, new methods of pasteurization are being developed to make shellfish safer than ever before. Increased world consumption fueled by renewed nutritional awareness has led to a variety of market forms to accommodate growing demand.

SHRIMP

STORAGE

With a shelf life of only a few days, fresh shrimp must be used immediately and kept as cold as possible (see Fig 3.1). Store shell-on shrimp buried in drainable crushed ice or in a covered container in the coldest part of the refrigerator. Avoid storing shell-off fresh shrimp in ice because it may burn the delicate flesh; instead store them in a covered container in the refrigerator and use immediately.

MARKET FORMS

The majority of shrimp are sold frozen or fresh. Because all shellfish deteriorate rapidly, fresh shrimp should always be iced and never sold or purchased above 40°F/4°C. Inspect for quality by smelling the shrimp; they should be briny without any stale, off, or fishy aroma. Shells must be intact and not at all slimy and the flesh should be firm without any yellowing. Frozen shrimp should be fully frozen without any sign of freezer burn or excess ice.

FIGURE 3.1 Shrimp should be received below 41°F/5°C. They should have a fresh, clean aroma and not be discolored or slimy.

Whole, head-on

Shrimp that are sold as they come out of the water are referred to as whole, head-on. They are available in various sizes, both fresh and frozen. Very authentic to many ethnic dishes, these shrimp have a dramatic effect on presentation. If they are peeled before service, the head can be used to make delicious shrimp stock. A disadvantage is that they are time consuming to fabricate and may not appeal to a fastidious diner. Because head-on shrimp is not in great demand, the frozen variety may remain frozen for unacceptable amounts of time, resulting in mealiness at the point where the head and body meet.

Green, headless

Also described as shell-on, this form is the most widely sold variety of shrimp in the world. Green refers to its raw state, not its color. It is available both fresh and frozen in many sizes or counts.

Peeled

Peeled shrimp are green headless shrimp with the shell and tail removed.

Peeled and deveined

These shrimp are sold fresh, frozen, cooked, and raw with the shell and vein removed and with the tail fin on or off.

Shell-on and cooked

These are sold fully cooked, with the shell and vein intact, ready to eat.

Butterflied

Raw shrimp that have been cut down the back and spread open. The shrimp should be fully intact with the vein removed.

Shrimp pieces

Typically sold frozen, these are shrimp with fewer than their normal six segments and are a very economical product suitable for a variety of cooking methods when whole shrimp are not required.

FROZEN SHRIMP PRODUCTS

IQF

The individually quick-frozen (IQF) process keeps each shrimp separated in the packaging and easy to remove as needed. IQF shrimp are available green or peeled and deveined, and can be thawed more rapidly than frozen blocks. The IQF process passes frozen shrimp through water before being quickly frozen again to form a protective glaze.

Blocks

Blocks are the standard packaging unit for most commercial kitchens. Sold in specific weights, they are wrapped in polyethylene and packaged in paper boxes. Water is added to the packaging before they are frozen. Thawing of blocks is best done in the refrigerator or under cold running water.

COOKED PRODUCTS

Shell-on or shell-off, cooked

Available fresh or frozen, these have been cooked and are sold shell-on or shell-off.

Breaded

A large variety of breaded shrimp products are available on both the retail and commercial market. Most are breaded in the raw state and then frozen. Various coatings and flavor profiles include seasoned breadcrumbs and batter. Sizes range from small to colossal. Formed products made from minced shrimp and even surimi are also available in a variety of shapes and sizes at greatly reduced prices. Breaded shrimp must be baked or fried in their frozen state or they will overcook and become soggy.

CONSISTENCY

With the high price of shrimp, it is important that you receive the correct count that you pay for. This can be easily calculated by establishing a uniformity ratio of a selected box of shrimp. Count 10

UNDERSTANDING SHRIMP COUNTS AND SIZES

Shrimp are sold in sizes or counts that are expressed in numbers per pound or kilogram (see Fig 3.2). As an example, shrimp marketed as 16/20 indicate the average number of shrimp per pound. Shrimp larger than 16/20 are called "U" meaning under; for example, U 15 or U 10 denote that there are under 15 or 10 shrimp per pound. The "U" designation starts at U/2 denoting shrimp with an average weight of 8 oz each. As the table on page 22 indicates, the average size of commercial warm-water shrimp ranges from U/10 to 61/70 shrimp per pound, from extra colossal to extra small. Cold-water shrimp are available in even smaller sizes ranging from 150/250 to 250/300. Price per pound is determined by the size; the larger the shrimp, the more expensive it will be.

Counts also vary between peeled and cooked shrimp; the actual number of shrimp in the package will be listed as the finished count.

FIGURE 3.2 Shrimp sizes from left to right: 31/40; 31-35, 21-25, 16-20; U10.

UNDERSTANDING SHRIMP SIZES AND COUNTS

SHRIMP SIZE	COUNT PER POUND GREEN, HEAD-OFF	COUNT IN A 5-LB/2.27 KG BOX
Extra colossal	Under 10	40–49
Colossal	Under 12 or Under 15	(U12) 50–59 (U15) 60–74
Extra jumbo	16/20	75–97
Jumbo	21/25	98–120
Extra large	26/30	121–145
Large	31/35	121–145
Medium large	36/40	174–190
Medium	41/50	191–249
Small	51/60	241–290
Extra small	61/70	291–340

percent of the largest shrimp and weigh them. Do the same with 10 percent of the smallest shrimp. Divide the weight of the largest by the weight of the smallest. The lower the ratio, the more consistent the count is. A uniform ratio of 1.0 indicates that they are all the same size. Because all shrimp differ in size, a range of 1.25 to 1.75 is more likely. When purchasing large quantities, periodic uniform ratio checks are important to ensure consistent portion size and food cost.

MOLLUSKS

PURCHASING GUIDELINES

Like most shellfish, mollusks are best purchased live. Certain varieties such as clams, oysters, and mussels can live out of water and hold up well to transportation and handling. Warm-water species such as the queen conch are typically very perishable and must be shucked and immediately eaten or frozen. Scallops have a relatively long shelf life out of the shell and typically are harvested, shucked, and refrigerated for transport. Fresh scallops in the shell with their roe intact are available at a premium price from specialized purveyors. When purchasing live mollusks they should always be in a net or open bag or they will suffocate. Inspect each one, discarding those that are cracked or completely open. Those with partially open shells can be tapped for signs of life; if they close up or move, they are still alive (see Fig 3.3b). Discard those that do not yield to pressure. Keep all mollusks refrigerated at between 34° to 38°F/1° to 3°C covered with a damp towel or seaweed; do not store them in ice (see Fig 3.3a). Shelf life varies depending on when they were harvested, but in general they should be consumed as quickly as possible; avoid holding them for more than several days.

Certified shellfish

Shipment of mollusks must be from a certified interstate source or approved state dealer, and all live products must be properly tagged. Any approved shipping lot that is not tagged is considered illegal and subject to seizure by government agencies. Shellfish tags contain specific information about the product that can be used should food poisoning occur from mollusks harvested from contaminated waters. Tags must remain attached until the container is empty, at which time the tags should be dated and kept on file for 90 days. Keep all mollusks separate and do not mix batches during storage. Containers of shucked shellfish must also have appropriate stickers or labels that show the name, certification number, and address of the packer. Containers of less than one-half gallon must also list the name of the product, the shucked date, and the sell-by date.

All shellfish should be received from a clean, cold delivery truck and are best stored between 34° to 38°F/1° to 3°C. Live shellfish should have a mild clean aroma and acceptable appearance. Shucked shellfish or mussels, clams, or oyster meats should smell briny, have clear liquor with an acceptable meat-to-liquid ratio, and be devoid of shells or sand. A FIFO (first in, first out) system should be practiced with all highly perishable foods based on quality characteristics and date of receipt.

SAMPLE SHELLFISH TAG:

Address New England Shellfish Co.	Certification # MA 1234	
Original shipper's certificate # xxxxx	If different from above xxxxxxxx	
Harvest date: 1/10/2007	Shipping date:	
Type of shellfish: oysters _____ mussels _____ soft clams _____		
Quantity of shellfish: bushel _____ pounds _____ count _____		
This tag must remain attached until container is empty and then kept on file for 90 days.		
To:	Reshipping certification #	Date of reshipping

FROZEN MOLLUSKS

Most mollusks can be purchased frozen in a variety of styles, forms, and packaging. Conch are sized and sold frozen in 5-pound boxes. Mussels, clams, and oysters are purchased individually frozen or frozen on the half shell. Abalone, readily available farm-raised, can be shipped live or frozen and vacuum-packed with individual weights of 1 ounce or greater. Scallops are available IQF or in vacuum packaging; the roe from the sea urchin is also available vacuum packed and frozen.

Breaded products

A variety of convenience-based frozen prebreaded and stuffed shellfish products are available on the commercial and retail level, sold by the box, case, or pound. Most should be cooked frozen to maintain their quality, flavor, and crisp texture. Some are breaded and frozen raw, whereas others are fully cooked.

FIGURE 3.3a Store live clams in the refrigerator covered with a damp cloth or seaweed. Cook or serve as soon after receiving them as possible.

FIGURE 3.3b Before preparing, tap any open clams. If they are alive, they will immediately close. Those that do not close are dead and should be discarded.

CLAMS

Clams are available year-round and come in a variety of market forms. Live clams are able to live out of water for a short period of time provided they are in a cold environment and have access to fresh air. Fresh-shucked clam meat comes in assorted sizes and forms and should be used as soon after receiving as possible. Make sure the tubs are free of noticeable shells and meet your specifications. Frozen whole bodies are available IQF or on the half shell. Other market forms include steaks, strips, chopped, and minced, as well as juice. Value-added products such as prestuffed shells and cakes are also available.

Because they are marketed under a variety of regional names which also differ based on the size within the same species, clam terminology is confusing. Information regarding individual species can be found in the identification chapter.

Clam counts (approximately)

Little necks	10 to 15 per pound
Topnecks	6 to 8 per pound
Cherrystones	3 to 5 per pound
Chowder clams	1 to 2 per pound
Soft-shell clams	12 to 15 per pound
Manila clams	20 per pound

Market forms
- Live, sold by the bushel, counts vary by species
- Fresh in tubs
- Chopped
- Frozen in blocks, whole, or in strips

- IQF, on the half shell
- Breaded strips and bellies
- Canned
- Smoked
- Chowder
- Juice
- Ground
- Farm-raised

MUSSELS

Like clams, mussels are available live year-round and can be kept for a short period of time in a refrigerator. As with all shellfish, avoid putting them in a sealed container or they will suffocate. Mussel shells will periodically open, so tap them first to determine whether they are still alive (see Fig 3.4a). If no motion is detected, discard the mussel.

Wild and farmed-raised varieties are available with cultivated mussels ranging in size from 2 to 3 inches. Modern aquaculture techniques ensure cleaner mussels with a higher meat ratio.

Market forms
- Live
- Smoked
- IQF, whole or on the half shell
- Canned
- Smoked
- Pickled
- Breaded
- Cooked and marinated in various sauces

OYSTERS

Oysters have many of the same shipping and storage requirements as clams and mussels. They are best purchased live with tightly closed shells and should be stored in the refrigerator, for no more than 48 hours, in a container or pan that allows for air circulation (see Fig 3.5). In many areas, live oysters are graded by shell shape and size and not for the meat quality and flavor, which differs between species and location. Oysters are typically purchased by the bushel which, depending on the species, contains approximately 100 to 300 oysters. Weighing about 75 pounds, a bushel will shuck down to only about a gallon of meat. On the retail level, oysters are purchased by the dozen or individually. To make them easier to open, pack them in ice or place them in the freezer for an hour; this

FIGURE **3.4a** Store mussels in the refrigerator covered with a damp cloth or seaweed. Cook or serve as soon after receiving them as possible. If the mussel is open, tap it; discard it if it does not move or close.

FIGURE **3.4b** Mussels have a beard that should be removed by pulling it back toward the hinge end of the shell. These threads are what the mussel uses to anchor itself to its living space. Cultured mussels are much cleaner than the wild-harvested varieties.

FIGURE **3.5** Store live oysters in the refrigerator covered with a damp cloth or seaweed. Cook or serve as soon after receiving as possible.

loosens the shell and makes it easier to open. Fresh oyster meat can be frozen in its own liquid for several months.

Pasteurized oysters

Vibrio vulnificus is a harmful bacterium that can be found in oysters. Over the past several decades, new technologies have been developed to reduce these naturally occurring bacteria to nondetectable levels in raw oysters. One of the first pasteurization methods after canning was the process of IQF. Although efficient at reducing or eliminating the bacteria, the freezing altered the physical characteristics enough to be questioned as a raw product. Other methods were developed, including low-heat pasteurization and a hydrostatic high-pressure system that introduces 45,000 psi of pressure into a controlled environment. This high-pressure treatment has the added bonus of opening the shell, eliminating the need for manual shucking. As the technology becomes more cost efficient with time, opening oysters with this technique will be increasingly prevalent.

Oyster counts

Standards: 200 to 300 per bushel
Selects: 100 to 200 per bushel
Extra selects: fewer than 100 per bushel

Market forms

- Live, singles or clusters, bushels
- IQF, on the half shell
- Smoked
- Shucked meats
- Breaded, frozen
- Tubs
- Gold band (pre-opened using pressurization)
- Farm-raised

SCALLOPS

Unlike clams and oysters, scallops cannot keep their shell tightly shut and once removed from the water they dry up and die, necessitating that they be shucked, eviscerated, and packaged promptly. Fresh intact scallops are available in the shell but must be chilled immediately and shipped in appropriate containers as expeditiously as possible.

FIGURE 3.6 Scallops in the shell should be received in a mesh bag or sack with an identification tag, which should be kept on file for 90 days after consumption. They should arrive alive and move when touched. If the scallops are not alive, they should be received below 41°F/5°C and smell sweet and briny.

Fresh scallops should have a shiny and glistening appearance and smell sweet and briny, not fishy and stale (see Fig 3.6). Their color will vary depending on species and location, but they are generally off-white to light tan with shades of orange, displaying a translucent quality. Aside from the shell, scallops are nearly 100 percent edible. During the shucking process, the viscera and foot are removed, but the orange-colored coral can be left on, adding to the flavor and presentation.

Wet and dry scallops

Because scallops dry out quickly, purveyors oftentimes soak them in a water bath containing sodium tripolyphosphate (see Fig 3.7). This process enables the scallop to retain up to 25 percent moisture, brightening its color, extending its shelf life, and plumping up its appearance. Unfortunately, this soaking makes them extremely difficult to cook; they become rubbery and their delicate nutty sweet flavor is compromised. Wet scallops look unnaturally white and plump, and although invariably cheaper, they are not worth the lower price because of their culinary shortcomings. Their longer shelf life means they are not as fresh.

FIGURE 3.7 The scallops on the left have been soaked in sodium tripolyphosphate. They are plump, white and very moist. The scallops on the right have not been soaked. They are off-white in color, slightly dry looking and droopy.

Dry scallops come directly from the shell and are not soaked in a water solution. They look dry although they will glisten and appear shiny. Dry scallops will not be as plump and will appear to sag, especially in the larger varieties. Their color will be off-white, light brown, to orange. In view of their high cost, it is important to understand the differences; questioning the purveyor will ensure a quality product.

Scallop counts

Although scallops are sold by the pound, they are sized and sorted similar to shrimp with the largest commanding the highest price at market. Sea scallops are the biggest of all the varieties, bay scallops are the smallest. A count of U/10 means there are fewer than 10 large sea scallops per pound with an average weight of approximately 1.6 ounces each.

U/10	Colossal sea scallops per pound
10/20	Jumbo sea scallops per pound
20/30	Large sea scallops per pound
80/120	Bay scallops per pound
150/250	Calico bay scallops per pound

Market forms

- Fresh in the shell
- Fresh shucked
- Dry or wet packed
- Tubs
- IQF, or block frozen
- Canned
- Smoked
- Breaded
- Farm-raised

ABALONE

Because of overharvesting and their slow growth and reproduction rates, abalone has been deemed unsustainable and commercial harvesting is illegal in the United States. A wide range of farm-raised products is available from a variety of sources in the Pacific Northwest and Canada. Wild Mexican stock is said to be sustainable. Because the

FIGURE 3.8 Abalone

abalone has only a flat top shell and is more like a snail than a clam, it is sensitive to environmental changes, making it difficult to store them live. For shipping, live abalones are placed in plastic bags that have been filled with oxygen, which will keep them alive for up to 48 hours. The oxygen-rich bags are packed in Styrofoam shipping containers with suitable ice packs. The best way to store live abalone is in a salt-water tank. During transport they can lose up to 15 percent of their body weight in water, decreasing yield and increasing cost. This loss of moisture can be recovered once the abalone is placed in a salt-water tank; if none is available, live product should be immediately removed from the shell and consumed. If they will not be used immediately, consider purchasing frozen product.

Abalone is a premium item that is sold by the piece or the pound. Individual live sashimi-grade abalone in the shell can cost over $30 a pound, and frozen large steaks sell for as much as $100 a pound.

Market forms
- Live farmed red abalone in shell
  - Small: approximately 3-inch shell length
  - Medium: approximately 4-inch shell length
  - Large: approximately 6-inch shell length
- Frozen farmed or wild Mexican whole abalone
  - Sizes vary dependent on producer
- Tenderized abalone steaks
- Canned
- Dried

CONCH

Easily gathered by wading into shallow water, or free diving to the ocean bottom, conch is in danger of being overfished. Sustainable farm-raised varieties are available from Turks and Caicos Islands in the British West Indies.

Market forms
- Wild variety in frozen 5-pound blocks
- IQF
- Fresh, farm-raised in a variety of sizes
- Canned

SEA URCHINS

Resembling a pincushion, the edible roe is considered a delicacy in Japanese cuisine where it is known as *uni* (see Fig 3.9). Each urchin contains five delicate lobes or sacs that must be carefully removed from inside the round shell. To remove the edible gonads or roe, as it is called, hold the shell in the palm of your hand, with the hole on top, cut the top third of the shell off with scissors, and remove the orange-colored roe. Roe is graded based on its

FIGURE 3.9 Sea urchin roe can be obtained by removing it from the sea urchin or by purchasing it from specialty purveyors. In Japanese, it is known as *uni* and is carefully packaged on wooden trays.

freshness, quality of color, and texture. The best grade is bright yellow or gold with a very firm texture and sweet flavor. Middle grade has a more muted yellow appearance, is softer in texture, and not as sweet. The lowest grade, referred to as *vana* in Japanese, is dark in color, very soft, and is often further processed by salting. Sea urchins can be purchased live and will keep refrigerated for up to 48 hours depending on shipping and transportation. They are packed in Styrofoam containers at or around 35°F/2°C. As with all shellfish, do not seal them in a container.

Market forms
- Live
- Frozen
- Baked and frozen
- Steamed
- Salted
- Paste form
- Canned

AMERICAN OR MAINE LOBSTERS

American or Maine lobsters have a great advantage of being able to live for several days out of water as long as they are in a cold, moist environment. They are best cooked live; their meat quickly spoils once they are dead. Because the viscera that breaks down the meat is found in the thorax, removing the lobster's "head" will slow down the deterioration process. Live lobsters are shipped in Styrofoam boxes containing ice packs and seaweed or wet newspapers. Dead lobsters can be identified by a limp tail and lack of reaction when picked up or placed in boiling water. Do not purchase them.

Storage
Lobsters should always be purchased live, cooked, or frozen (see Fig 3.10). Store live lobsters in aerated salt-water holding tanks or in the refrigerator covered with damp newspaper or seaweed for as little time as possible. Do not pack live lobsters in ice and do not submerge them in fresh water; this will kill them. Monitor live lobsters for lethargic signs and cook them as soon after receiving as possible.

Sizes and terminology
Legal size for a lobster is between 3 1/4 and 5 inches from the rear of the eye socket down to the main body shell; this area is called the

FIGURE **3.10** Lobsters should always be received alive or frozen solid, often glazed. Alive, they should be packed in Styrofoam boxes with ice packs and damp newspaper or seaweed. Do not purchase lifeless lobsters.

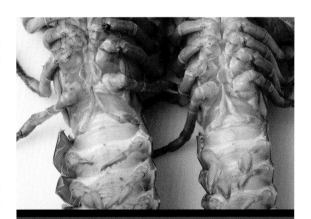

FIGURE **3.11** Because female lobsters contain an edible coral or roe, it may be important to determine the gender of lobsters (female on left and male on right). Turn the lobster onto its back: the male has a narrow abdomen whereas the female's abdomen is broader. Below the walking legs are two small appendages pointing up toward the body. These are featherlike and crisscrossed in females and somewhat larger and hard in males.

carapace. Each whole lobster yields approximately 20 percent of its total body weight in edible meat, and is marketed in a variety of sizes.

Chickens	approximately 1 pound
Quarters	approximately 1 1/2 pounds
Select	approximately 1 1/2–2 1/2 pounds
Jumbo	more than 2 1/2 pounds
Culls	whole lobsters with only one claw
Soft shells	lobsters that have recently molted; their shell will not be as hard and the meat will be spongy and have lower yield. Avoid soft-shell lobsters.

SPINY LOBSTER

Found in the southwestern Atlantic, California, South Africa, Australia, and the Mediterranean, the spiny lobster is clawless and is known for its hard shell and tail meat. It is normally sold frozen. From a culinary perspective, the meat has a coarser texture and is not as sweet as its New England cousin. Water is often pumped between the shell and meat for glazing purposes, which also increases the price per pound. Many varieties are sold in tail form including the rock lobster, slipper lobster, and Spanish lobster.

Market forms of lobster

- Live
- Frozen whole cooked
  — Cooked and vacuum packed; these are sometimes glazed or packed with brine
- Frozen raw tails
  — Individually vacuum-packed tails, IQF, and sold in a variety of weights
- Fresh or frozen lobster meat
  — A wide variety of meat combinations including tail, claw, knuckle, and custom blends are available fresh or frozen, vacuum packed or canned
- Lobster bodies
  — Available frozen, the bodies are used to make stock or sauce
- Lobster tomalley and eggs
  — The tomalley, or liver, as well as the roe, or eggs, are available from specialty purveyors

CRABS

Crabs are able to breathe out of water and can survive in the refrigerator covered with damp newspapers or seaweed for up to 48 hours (see Fig 3.12). Purchasing live crabs will ensure a quality product, but the skill and labor involved to remove the meat is an important cost factor. Because of the meat's jelly-like consistency when raw, it must be cooked prior to removal. Even when cooked it is difficult and time consuming to free it from the many shells and body chambers.

FIGURE 3.12 Store live crabs in the refrigerator covered with a damp cloth or seaweed. Cook them as soon after receiving as possible.

A variety of crabs is fished on both coasts of the United States and is available year-round in an assortment of market forms.

Determining quality

Cooked whole crabs should have a clean, moist shell with the legs intact; the weight should match the body size. If a crab has not yet filled out its shell from the molting stage, it will have a lesser yield. As with all shellfish, a sweet briny aroma is an indicator of quality.

Pasteurized or canned crabmeat should be well-sealed, free from noticeable shells, and look and smell clean. Frozen meat should not be dried out or exhibit freezer burn. Lump meat should fall within the standard of purchases, especially in the prime jumbo lump form. Sort through a sampling of the containers to determine that you are getting full value.

Market forms
- Live blue crabs
- Soft-shell crabs, fresh or frozen
- Blue crab meat
  — Jumbo lump
  — Back fin lump
  — Claw meat
- Jonah crab, available live, whole cooked, cooked and pasteurized, cocktail claws, whole claw, and arm sections
- King crabs, whole cooked legs, split legs, claws, pasteurized, or canned meat
- Snow crab, legs, split legs, leg clusters, claws, sections, pasteurized picked meat
- Dungeness crabs, live, cooked, or whole frozen. Picked meat is also available pasteurized or canned.
- Stone crab
  — Cooked fresh or frozen claws
- Frozen products
  — Crab cakes, crab puffs, imitation crab sticks (surimi)

SQUID

Squid can be purchased in a variety of market forms depending on freshness and convenience. Piles of fresh inky squid pulled out of local waters are available at fish markets around the world, but most consumers prefer them to be precleaned. Whole squid contain the head, tentacles, ink sac, mantle, and pen or quill, which is the simple internal shell or cartilage. Fresh squid should smell briny, without any off or fishy aroma. The bodies will be various shades of gray with small dark spots. Fresh squid should be cleaned immediately and refrigerated; it also freezes very well. Frozen squid, fully cleaned, comes in a variety of forms and may have been soaked in a whitening solution. These can easily be identified by their unnaturally white appearance.

Market forms
- Whole, fresh
- Whole, fresh cleaned
- Frozen
  — Whole
  — Cleaned
  — Rings
  — Tubes
  — Tentacles
- Ink
- Breaded
- Stuffed
- Dried
- Canned
- Smoked

FIN FISH IDENTIFICATION

ANCHOVY

- European (*Engraulis encrasicolus*)
- Northern (*Engraulis mordax*)

Anchovies are prized throughout the world for their unique flavor and preservation qualities. Beautifully iridescent gray and blue-green, they average 6 to 9 inches long and have a short blunt nose, a wide jaw, and sharp teeth. They are found in the southern and northeast Atlantic, and throughout the Mediterranean. The most common species from a culinary and production standpoint is the European anchovy (*Engraulis encrasicolus*). The meat is rich in omega-3 fatty acids, dark in color, and eaten fresh, smoked, and cured.

Along the Pacific coast, deep ocean currents collide with coastal cliffs, bringing cold water rich in plankton and nutrients to the surface. This large food source attracts the anchovy and other marine life, making it one of the most productive fishing and biodiverse areas in the region. Off the coast of Peru, millions of tons of anchovies are harvested and used for an ever-increasing fish meal and fish oil industry. Important for their nutrient-rich qualities, including high amounts of omega-3 fatty acids, they are also used in the production of livestock and aquaculture feed.

In the United States, fresh anchovies do not have a great commercial value, and most are sold in cans or paste. Historically, they were an important ingredient in the ancient Mediterranean sauce called garum, which was one of the first sauces made and shipped in commercial quantities. An important species throughout Asia, it is used in the production of many fish sauces, and is an ingredient in Worcestershire sauce.

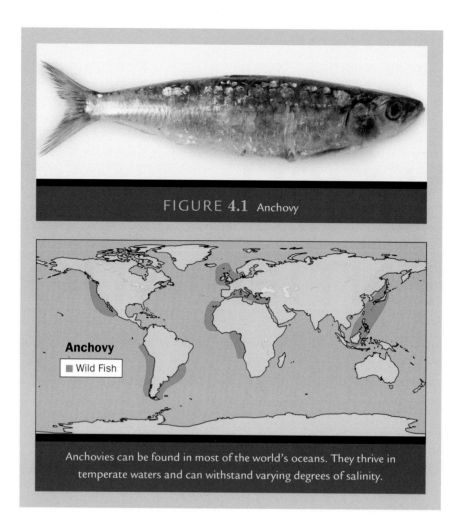

FIGURE 4.1 Anchovy

Anchovies can be found in most of the world's oceans. They thrive in temperate waters and can withstand varying degrees of salinity.

QUALITY CHARACTERISTICS

Most of the "fishy" flavor of the anchovy comes from the curing process. When eaten fresh they are rich and moist, but contain innumerable pin bones.

Available seasonally in the United States, they are typically sold head-off. Heretofore underutilized in fresh form, anchovies are increasingly popular with culinarians who appreciate the versatility and low cost.

PREPARATION METHODS

- Bake
- Deep fry
- Pan fry
- Broil
- Grill

SEASONAL AVAILABILITY

January to August

ANCHOVY (3 OZ SERVING)

Water: 62.36 g
Calories: 111
Total fat: 4.11 g
Saturated fat: 1.09 g
Omega-3: 1.747 g
Cholesterol: 51 mg
Protein: 17.30 g
Sodium: 88 mg

ARCTIC CHAR (*SALVELINUS ALPINUS*)

A relative of trout and salmon, arctic char is one of the most northerly fish of Europe, Asia, and North America. Ranging in size from 3 to 8 pounds, they are found in both lakes and oceans, and are extensively farm raised. They are a beautiful speckled fish with varying degrees of darker colors on the back and an upwardly protruding lower mouth. The belly color always fades into a lighter solid, which along with the mouth streamlines the fish. The male pelvic, anal, and pectoral fins are red. All fish have a smaller adipose fin located behind the dorsal fin. These fish are a good choice for aquaculture because they have the ability to withstand cold water conditions and consume relatively little feed. It is feared that in the future, global warming and increased land use will endanger this heretofore unspoiled fish. Generally, there are concerns of depletion of other species used for fish meal. However, arctic char consume relatively little feed; it takes as little as 1 to 2 pounds of feed to produce 1 pound of fish.

QUALITY CHARACTERISTICS

Mild in flavor and closer to the taste of trout than salmon, the meat has a small flake and is adequately fatty, holding up well to many cooking methods including curing and smoking. Wild species have a wide range of colors, textures, and flavors depending on environmental conditions. Because of the distance to market, much of the product is frozen, but fresh is available. Purchase fresh fish well chilled with clear eyes and gills that are intact and red. The fish should appear colorful, the flesh should be firm, and there should be no noticeable odor.

Because it has a high fat content and is milder than salmon in flavor, Arctic char is very versatile in the kitchen. It is delicate enough to poach, but firm and flavorful enough to smoke or make into a mousseline. Remove the skin before serving.

PREPARATION METHODS

- Bake
- Broil
- Grill
- Poach
- Sauté
- Smoke

FIGURE 4.2 Arctic Char

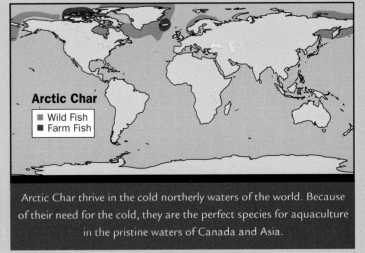

Arctic Char thrive in the cold northerly waters of the world. Because of their need for the cold, they are the perfect species for aquaculture in the pristine waters of Canada and Asia.

SEASONAL AVAILABILITY

Wild and farm-raised varieties available year-round.

ARCTIC CHAR (3 OZ SERVING)

Calories: 139 Total fat: 6.72 g Saturated fat: 1.44 g
Cholesterol: 23 mg Protein: 18.19 g Sodium: 55 mg

BARRAMUNDI (*LATES CALCARIFER*)

Barramundi is a species of bass inhabiting rivers and estuaries in Australia, and it has had great success being farm-raised. Meaning "large scales" in Aboriginal language, it is shiny gray in color and can grow to more than 100 pounds in the wild. Barramundi are typically available farm-raised between 1 and 3 pounds, are raised in closed indoor systems, free of hormones and antibiotics, and consume protein-rich soy meal. This fish has been served at a White House state dinner honoring the Australian Prime Minister.

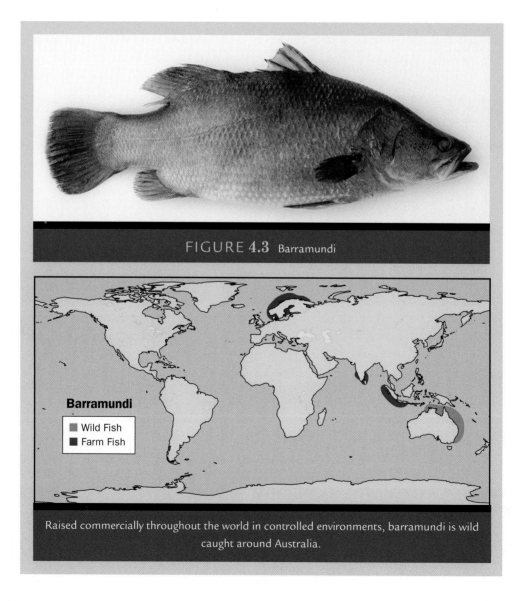

FIGURE 4.3 Barramundi

Raised commercially throughout the world in controlled environments, barramundi is wild caught around Australia.

QUALITY CHARACTERISTICS

Barramundi meets consumers' needs with its mild flavor and firmness. Even large-scale chains such as Wal-Mart are interested in the sustainability of species such as Barramundi.

PREPARATION METHODS

- Bake
- Fry
- Sauté
- Broil
- Poach
- Steam

SEASONAL AVAILABILITY

Nearly all of the U.S. supply is farm-raised and available year-round.

BASS, CHILEAN SEA (*DISSOSTICHUS ELEGINOIDES*)

The actual name for this species is the Patagonian tooth fish, a member of the Nototheniidae family. Its name was changed when it was first imported into the United States from Chile. This new name is a prime example of how marketing is used to promote and drive sales of a previously unknown species. Inhabiting the southern parts of South America and Africa, it can grow to several hundred pounds with market size ranging from 20 to 40 pounds. They have long, mottled gray bodies and snow-white meat. Because they are caught in such remote areas, the bulk of the fish is frozen at sea. Increasingly, it is being flown out of Santiago, Chile to metropolitan areas around the world.

Although not considered endangered by the U.S. government, there are a lot of fish taken illegally, which is especially unfortunate because of its slow growth rate. The Commission for the Conservation of Antarctic Marine Living Resource is a 24-country

FIGURE 4.4 Bass, Chilean Sea

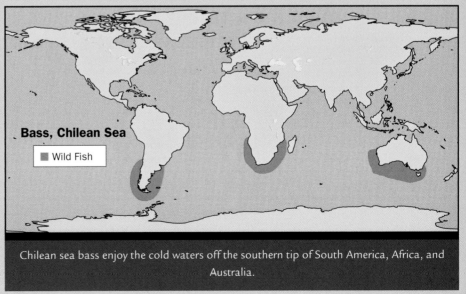

Chilean sea bass enjoy the cold waters off the southern tip of South America, Africa, and Australia.

body that institutes catch limits and management practices. In the United States, U.S. Customs and National Oceanic Atmospheric Administration (NOAA) allows imports into U.S. markets only with a valid dealer permit, and the Marine Stewardship Commission, a nonprofit organization that promotes responsible fishing practices, has recognized specific fishing grounds as sustainable. Despite regulations, and until chefs and consumers insist on certified fish, it will continue to be sought-after worldwide, especially in Japan where its rarity, flavor, and texture add to its mystique.

QUALITY CHARACTERISTICS

Purchase this fish knowing its origins and insist on documentation from your supplier. Fresh fish are typically sold head-off and should be moist, firm, and smell like the ocean. Avoid fish that are dried and yellowing or are not packed correctly in crushed ice. If purchasing frozen or refreshed fillets, they should be bright white with no noticeable flake separation. Thawing should be done slowly at temperatures between 35° and 40°F/2°C and 4°C.

From a culinary standpoint, the only deterrent to this beautiful fish is its high price and questionable sourcing. Buttery in texture and flavor, the large fillets or steaks from this fish contrast nicely with a wide variety of cooking methods and accompaniments. Bright white when cooked, the meat has a high enough fat content to keep it moist but not enough to negatively affect the rich flavor. Because this is a premium fish, treat it simply with accompaniments that contrast its flavor and texture.

PREPARATION METHODS

- Bake
- Grill
- Sauté
- Broil
- Poach
- Steam

SEASONAL AVAILABILITY

Available fresh from April to July and frozen year-round.

BASS, CHILEAN SEA (3 OZ SERVING)

Water: 66.53 g	Calories: 82	Total fat: 1.70 g	Saturated fat: 0.434 g
Omega-3: 0.822 g	Cholesterol: 35 mg	Protein: 15.67 g	Sodium: 58 mg

BASS, HYBRID (*MORONE CHRYSOPS-M. SAXATILIS*)

As its name implies, this hybrid was developed by crossing the striped bass and the white bass, fish that inhabit both fresh and brackish waters. The resulting species combines the great flavor of the wild striped bass with the fast-growing, resilient qualities of the white bass. With a market size between 1 and 3 pounds, this fish is meatier than its wild cousins and is farmed in a variety of methods including ponds and indoor tanks. They are also an excellent fish for aquaponics, which is the process of using the by-products or waste from one species, the hybrid bass, to aid in the production of another item in the same proximity. As an example, the fish can be raised in indoor tanks and their waste and water evaporation can be used to grow an array of plants, vegetables, and herbs.

The hybrid bass can be identified by its small mouth and broken longitudinal strip pattern. It is silvery blue to black, fading to white on the belly. Most fish are sold whole or as skin-on fillets, which aids in identification. Moderately firm in texture, the meat is mild in flavor with a faint earthy taste. When cooked, the meat is an off-white color. Because of its small size, it is an ideal fish to serve whole.

QUALITY CHARACTERISTICS

Typically sold whole, the fish should be packed in ice and have a clean nonfishy aroma. If purchased as fillets, they should be skin-on to aid in identification. The eyes should be bulging and the flesh firm. The scales should be intact and the overall appearance must be moist and not dried out, which is a sign of improper handling. Hybrid bass are also sold alive and are popular in Asian markets. Many farms are in operation throughout the Mid-Atlantic and southern part of the United States, ensuring a fresh and quality product.

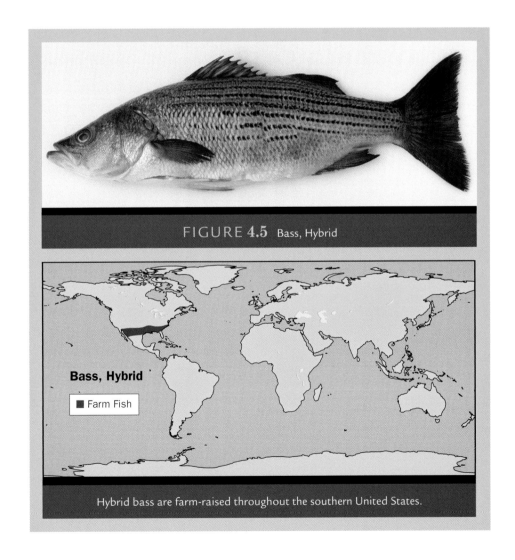

FIGURE 4.5 Bass, Hybrid

Hybrid bass are farm-raised throughout the southern United States.

Hybrid black sea bass has a moderately firm texture and is an excellent small fish to serve whole. The skin is beautiful and flavorful when served crispy; be sure to remove all the scales before cooking.

PREPARATION METHODS

- Bake
- Deep fry
- Pan fry
- Sauté
- Stew
- Broil
- Grill
- Poach
- Steam

SEASONAL AVAILABILITY

Year-round

BASS, HYBRID (3.5 OZ SERVING)

Calories: 96　　Total Fat: 2.4 g　　Saturated Fat: 0 g　　Omega-3: 0.9 g
Cholesterol: 80 mg　　Protein: 17.6 g　　Sodium: 69.4 mg

Source: Seafood Handbook (Try-Foods International)

BASS, EUROPEAN SEA (*DICENTRARCHUS LABRAX*)

Also known as *loup de mer* in French and *branzino* in Italian, the European sea bass is found in the Eastern Atlantic and the Mediterranean Sea. It is also farmed in northern Europe and Greece. Most fish average 2 to 5 pounds. They are a light silvery blue-gray on the upper portion of the body and have an off-white belly. A lateral stripe runs the length of the fish; it has two dorsal fins. Due to high shipping costs, only a select number of U.S. restaurants serve the European sea bass.

QUALITY CHARACTERISTICS

Because the fish must be transported from European waters, it is important to examine it carefully for freshness. It should smell fresh and clean, and the gills should be intact and bright red. The fish should not appear dry and the scales should be firmly attached to the body.

The meat is highly prized throughout Europe for its delicate texture and mild lean flavor, which lends itself to cooking methods that introduce fat and moisture. In French cuisine it is typically sautéed.

PREPARATION METHODS

- Bake
- Deep fry
- Pan fry
- Sauté
- Stew
- Broil
- Grill
- Poach
- Steam

SEASONAL AVAILABILITY

Year-round

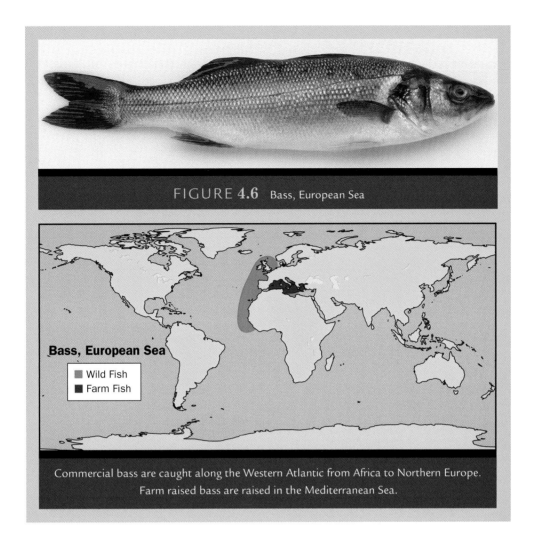

FIGURE 4.6 Bass, European Sea

Commercial bass are caught along the Western Atlantic from Africa to Northern Europe. Farm raised bass are raised in the Mediterranean Sea.

BASS, EUROPEAN (3 OZ SERVING)

Water: 64.31 g Calories: 97 Total fat: 3.14 g Saturated fat: 0.663 g
Omega-3: 0.649 g Cholesterol: 58 mg Protein: 16.03 g Sodium: 60 mg

BASS, BLACK SEA (*CENTROPRISTIS STRIATA*)

Also referred to as rock bass or blackfish, the black sea bass is sometimes misidentified as striped bass. Averaging 1-1/2 to 3 pounds, they inhabit the East Coast of the United States from New England to Florida and are caught using long line, trawl nets, and fish traps. The black sea bass can easily be identified by its shimmering black body and longitudinal white diamond markings, but like all fish, coloration is based on specific environmental circumstances. The dorsal fin is large and continuous, containing 10 sharp spines. The fish has a large mouth and large scales.

QUALITY CHARACTERISTICS

Moderately firm in texture, black sea bass is mild in taste, and its flesh very white when cooked. One of the few fish sold live, it is found in the tanks of many Asian restaurants along the East Coast. Once fabricated, fillets have a longer shelf life than most fish and are usually not marketed frozen. Gutted fish should not appear dry and the large scales should be firmly attached to the body. Because this fish is typically sold whole, when purchasing fillets, they should be skin-on to aid in identification. The skin is beautiful and flavorful when served crispy; be sure to remove all the scales before cooking.

PREPARATION METHODS

- Bake
- Deep fry
- Pan fry
- Sauté
- Stew
- Broil
- Grill
- Poach
- Steam

SEASONAL AVAILABILITY

Year-round

BASS, BLACK SEA (3 OZ SERVING)

Water: 66.53 g   Calories: 82   Total fat: 1.70 g   Saturated fat: 0.434 g
Omega-3: 0.822 g   Cholesterol: 35 mg   Protein: 15.67 g   Sodium: 58 mg

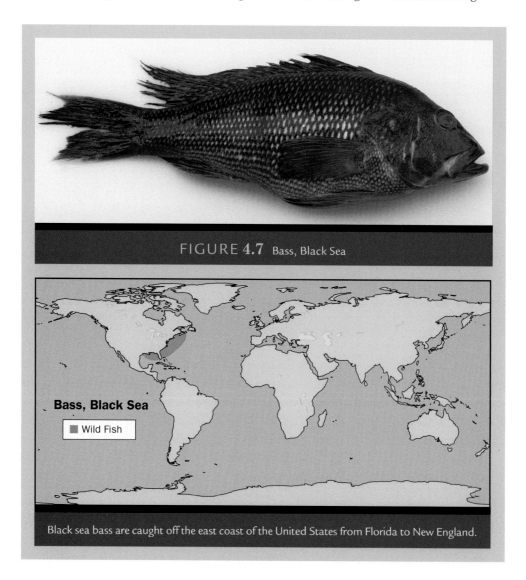

FIGURE 4.7 Bass, Black Sea

Black sea bass are caught off the east coast of the United States from Florida to New England.

BASS, STRIPED (*MORONE SAXATILIS*)

Striped bass inhabit the Atlantic coastline from the Gulf of Mexico north to the St. Lawrence River, with concentrations from Cape Cod to South Carolina. The Chesapeake Bay, Hudson River, and Delaware River are the three major spawning grounds; extensive conservation efforts in the 1990s have brought the species back to sustainable levels.

Stripers, as they are also called, are an anadromous species similar to salmon in that they live most of their life in salt water and migrate up fresh water rivers in the spring to spawn. Throughout the world, they have been introduced into fresh water lakes and reservoirs as game fish and as predators to control various unwanted species. In the late 1800s they were established in the Pacific Ocean and now can be found from Southern California to Washington State.

QUALITY CHARACTERISTICS

Silvery blue-gray in color, its consistent dark lateral stripes set it apart from its hybrid cousin whose lateral stripes are noticeably broken. Striped bass can weigh well over 50 pounds, but typical market size range between 5 and 15 pounds. Many chefs feel that the very large fish have a mealy texture, their potential of PCB (polychlorinated biphenyl) contamination should encourage consumers to check with local government

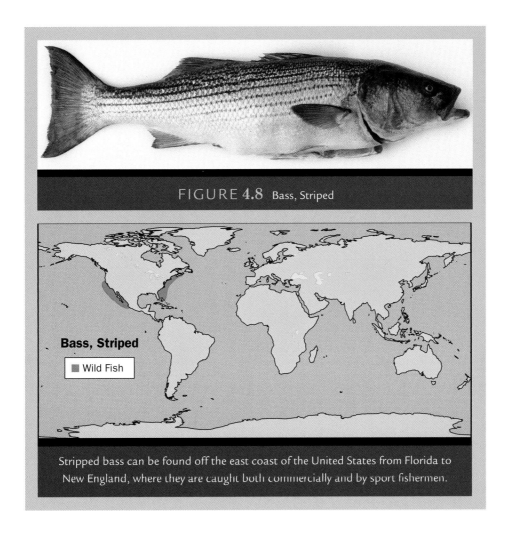

FIGURE 4.8 Bass, Striped

Stripped bass can be found off the east coast of the United States from Florida to New England, where they are caught both commercially and by sport fishermen.

agencies for more detailed information regarding repeated consumption. The meat is white in color, with a mild flavor and medium texture.

PREPARATION METHODS

- Bake
- Broil
- Fry
- Poach
- Raw
- Sauté
- Steam

SEASONAL AVAILABILITY

Year-round

BASS, ATLANTIC STRIPED (3 OZ SERVING)

Water: 67.34 g   Calories: 82   Total fat: 1.98 g   Saturated fat: 0.431 g
Omega-3: 0.822 g   Cholesterol: 68 mg   Protein: 15 g   Sodium: 59 mg

BLACKFISH, TAUTOG (*TAUTOGA ONITIS*)

Found from Canada south through the Carolinas, the blackfish averages about 3 pounds and is gray to black with mottled white spots on its body. Also referred to by the American Indian name *tautog*, they can be identified by their molar-like teeth and

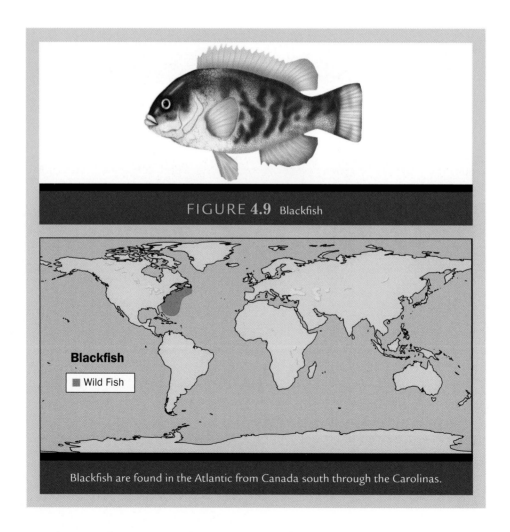

FIGURE 4.9 Blackfish

Blackfish are found in the Atlantic from Canada south through the Carolinas.

full lips. Caught mostly by party boats throughout New England, they are also landed by trawlers' gill nets and as bycatch in lobster traps.

QUALITY CHARACTERISTICS

Blackfish have a medium firm white meat that is mild in flavor and lends itself well to fish chowders.

PREPARATION METHODS

- Bake
- Fry
- Stew
- Broil
- Sauté

SEASONAL AVAILABILITY

Year-round

BLUEFISH (*POMATOMUS SALTATRIX*)

Enjoyed by anglers for its aggressive fight, this migratory fish ranges from New England to South America with average weights from 3 to 20 pounds. They are

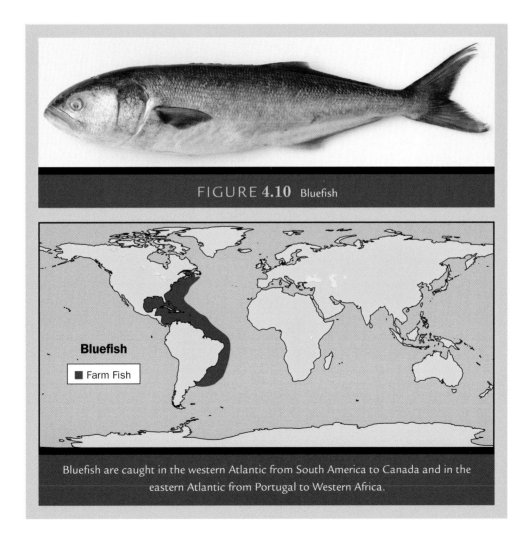

FIGURE 4.10 Bluefish

Bluefish are caught in the western Atlantic from South America to Canada and in the eastern Atlantic from Portugal to Western Africa.

a silvery blue-green on top with a silver gray-blue belly and a short dorsal fin. Bluefish have large mouths and jaws and very sharp teeth. Small fish are referred to as snappers and the large ones are known as horses.

QUALITY CHARACTERISTICS

Because of its high fat content bluefish have a short shelf life and must be iced and consumed quickly. Smaller fish are more delicate in flavor than large fish, which are oily and contain a blood line that should be removed. Fresh fish are typically available seasonally and larger fish have been known to have significant levels of PCBs. Gills should be intact and very red. The flesh should be firm and opaque to light brown. Fish should smell fresh.

PREPARATION METHODS

- Bake
- Broil
- Grill
- Sauté
- Smoke

SEASONAL AVAILABILITY

Year-round

BLUEFISH (3 OZ SERVING)

Water: 60.23 g Calories: 105 Total fat: 3.60 g Saturated fat: 0.778 g
Cholesterol: 50 mg Protein: 17.03 g Sodium: 51 mg

BREAM (*SPARUS AURATUS*)

A member of the porgy family, the sea bream is found throughout the Mediterranean where it is also farm-raised. It can be identified by a gold stripe between its eyes and steep head. Coloration ranges from silvery gray to blue with varying shades of purple around the belly and a lateral line from the gills to the tail. Market size averages between 1-1/2 and 5 pounds. It is highly sought after by European chefs where it is sold under the name of *daurade*. Beware of substitutes especially when purchasing fillets; it is best to purchase whole fish.

QUALITY CHARACTERISTICS

The meat is an off-white to red color, with a firm texture and sweet pronounced non-fishy flavor. Small fish are best cooked whole; the firm meat holds up well to moist heat cooking methods.

COOKING METHODS

- Bake
- Braise
- Broil
- Grill
- Poach
- Sauté
- Stew

SEASONAL AVAILABILITY

Year-round

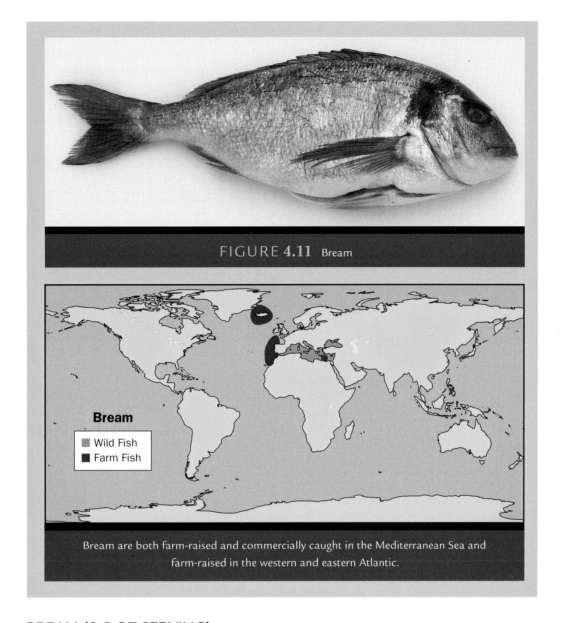

FIGURE 4.11 Bream

Bream are both farm-raised and commercially caught in the Mediterranean Sea and farm-raised in the western and eastern Atlantic.

BREAM (3.5 OZ SERVING)

Calories: 96   Total Fat: 1.9 g
Omega-3: 0.4 g   Protein: 19.7 g

Source: Seafood Handbook, their source Frimodt

CATFISH (*ICTALURUS PUNCTATUS*)

Once a muddy-tasting fish caught with rod and reel from fresh water lakes, the catfish is now a model of aquaculture success throughout the Mississippi delta and surrounding areas. Looking for new crops to supplement their soybeans and rice, farmers began raising channel catfish in ponds. Fast growing, this fish has the unprecedented ability to produce 1 pound of usable meat for every 2 pounds of vegetarian-based feed. Most are harvested between 1 and 2 pounds. Farming in this country is regulated by the Food and Drug Administration ensuring a consistent, low-cost quality product using sustainable techniques and practices. Although available whole, most catfish are sold in fresh or frozen fillets. The channel catfish can be found from the Great Lakes south to the Carolinas and Mexico. It has a forked tail and is varying shades of gray with black spots and characteristic whiskers.

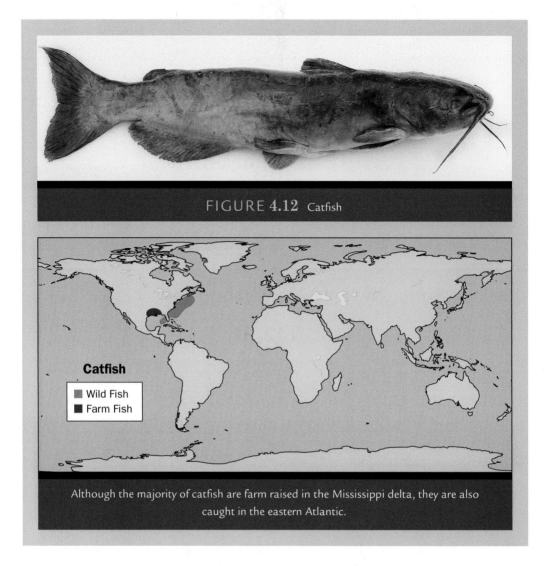

FIGURE 4.12 Catfish

Although the majority of catfish are farm raised in the Mississippi delta, they are also caught in the eastern Atlantic.

QUALITY CHARACTERISTICS

The meat of farm-raised fish is moist, mild, and sweet, and the texture is firm, almost dense. The fillets are varying shades of white and sometimes have light red undertones. Quality is normally outstanding but as with all farmed fish an imbalance of water conditions can lead to muddy flavors in the meat.

PREPARATION METHODS

- Bake
- Deep fry
- Sauté
- Broil
- Grill

SEASONAL AVAILABILITY

Year-round

CATFISH (FARMED) (3 OZ SERVING)

Water: 64.07 g	Calories: 115	Total fat: 6.45 g	Saturated fat: 1.503 g
Omega-3: 0.150 g	Cholesterol: 40 mg	Protein: 13.22 g	Sodium: 45 mg

COD (FAMILY: GADIDAE)

Commercial species:

- Atlantic
- Cusk
- Haddock
- Hake
- Pollock
- Whiting
- Pacific

When the ancient fishermen set out from northern Europe and the Mediterranean, they encountered a seemingly limitless supply of cod spawning close to shore in the shallows or in the slightly deeper waters offshore. Cod live in areas where warm and cold water currents converge. Off the coast of New England, the warm Gulf Stream flows by the colder Labrador Current, churning up the sea and creating an abundance of plankton, fish, and other marine life on which the cod feed. In the Pacific off Alaska, the cold arctic current and the warm Japanese currents support the Pacific species.

From Newfoundland to New England lie shallow shoals or banks at the edge of the Continental shelf. In the north lie the Grand banks and south off of Massachusetts is the Georges Bank. For centuries, these areas supplied much of the Western world with one of the most important commercial fish of all time. Extremely prolific, the female can produce tens of millions of eggs over her 20-year life. Unfortunately, overfishing and lack of controls have forced these sacred fishing grounds to close.

Codfish consists of many family groups and species, with the majority living in the northern hemisphere. Exceptions include a tropical water species, and the fresh water burbot, inhabiting lakes in New England, Alaska, and the northern United States and Canada. Cod is one of the most amazing fish of the world. It lends itself to all cooking methods and has a wonderful mild flavor. For centuries cod has been dried and salted and used as trading currency throughout the Caribbean, South America, the Mediterranean, and West Africa. It has always been a shelf-stable nutritious protein to the warmer climates of the Western world. In Scandinavia, lutefisk is made from dried cod that is soaked in a solution of lye and water until soft and jelly-like. Further soaking in fresh water draws out the lye before it is cooked and served with accompaniments such as bacon. In addition to its meat, the roe, cheeks, tongue, and air bladder are edible. Generations of children remember with horror the spoonful of cod liver oil which we now know is high in omega-3 fatty acids.

ATLANTIC COD (*GADUS MORHUA*)

The most commercially important of all cod species, the Atlantic cod is fished in the cold salty waters of the Northern Hemisphere using nets or trawlers, but the flesh of the line-caught fish is superior in quality. Cod are bottom-feeders with a diet consisting of crustaceans and other small marine life. For centuries they have been one of the most heavily fished species and are now strictly regulated. Cod are prone to infection from a small roundworm called a *nematode*, which can be identified by placing the fillet on a light box. Nematodes are not a health hazard if the fish is properly cooked or frozen.

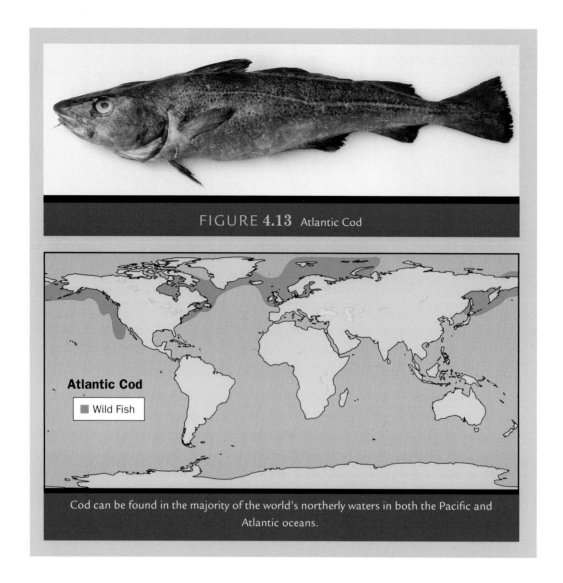

FIGURE 4.13 Atlantic Cod

Cod can be found in the majority of the world's northerly waters in both the Pacific and Atlantic oceans.

SCROD (*GADUS MORHUA*)

In New England, the term *scrod* refers to a small Atlantic codfish, pollock or haddock of about 2 pounds or less. From the Dutch word *schrood*, it means a cutoff piece or fillet.

IDENTIFICATION

Atlantic cod is speckled with brownish spots on a body of varying colorations of brown, green, yellow, and gray depending on water conditions and location. A pale white lateral line on the sides highlights the distinctive five fins. Its tail is squarer in shape without being forked, and a barbel that hangs from under the chin is used to stir up food on the ocean floor. Atlantic cod is the largest of the species and can grow to more than 20 feet and reach a weight of over 100 pounds. Most commercial fish are within the 6 to 20 pound range.

QUALITY CHARACTERISTICS

The flesh should be white and steaks should glisten, not appear dry. Thicker cuts from around the shoulder area are preferred for baking, broiling, grilling, pan frying, poaching, and sautéing. Tail pieces are superior for chowders and deep frying.

For centuries, the entire fillet has been salted and contemporary chefs are beginning to prepare "in house" cured cod, which has become versatile in today's kitchen. Cod has an exceptional mild flavor and texture, able to complement and contrast even the boldest ingredients. Bacon, cabbage, Asian and spicy flavors, saffron, and red wine reductions all work well with cod, as do beans and many wines high in minerals, oak, or green flavors. Cod cheeks, also available, are wonderful floured and sautéed, served with a reduction or beurre blanc sauce.

COD, ATLANTIC (3 OZ SERVING)

Water: 69.04 g	Calories: 70	Total fat: 0.57 g	Saturated fat: 0.111 g
Omega-3: 0.134 g	Cholesterol: 37 mg	Protein: 15.14 g	Sodium: 46 mg

CUSK (*BROSME BROSME*)

Sometimes misidentified as cod, whiting, and scrod. It has a cylindrical body with blotchy colorations of brown and gold on top and a lighter belly. The dorsal fin runs form the nape of the head to the caudal peduncle, and the anal fin begins after the belly and ends at the tail. The characteristic chin barbel also aids in identification. Trawled

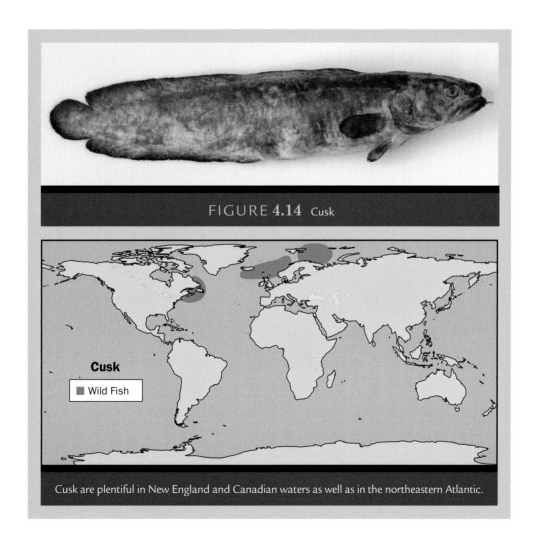

FIGURE 4.14 Cusk

Cusk are plentiful in New England and Canadian waters as well as in the northeastern Atlantic.

from Greenland to New Jersey; average size is between 10 and 20 pounds. Cusk is one of the firmest of the cod family and is mild in flavor. A line of bones runs down the center of the fillets. In addition to fresh and frozen, it is often salted and eaten as salt cod. In Spain it is known as *bacalao*.

CUSK (3.5 OZ SERVING)

Calories: 87   Total fat: 0.7 g   Cholesterol: 41 mg
Protein: 19 g   Sodium: 31 mg

Source: Seafood Handbook, their source USDA

HADDOCK (*MELANOGRAMMUS AEGLEFINU*)

Dark brown to gray skin, with two black spots on either side of the pectoral fin and a black lateral line on the sides; smaller and cheaper than cod, they range from 2 to 5 pounds. The meat is softer and less flaky than cod and has a more delicate flavor. Haddock is excellent battered and fried, or made into a mousselline. Finnan Haddie is a Scottish preparation in which the sides of haddock are dyed a golden color, then dried and smoked; they are sold in whole sides.

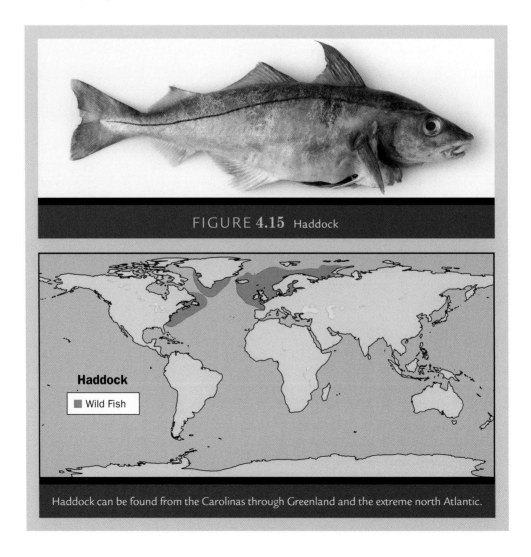

FIGURE 4.15 Haddock

Haddock can be found from the Carolinas through Greenland and the extreme north Atlantic.

HADDOCK (3 OZ SERVING)

Water: 67.93 g	Calories: 74	Total fat: 0.61 g	Saturated fat: 0.111 g
Omega-3: 0.202 g	Cholesterol: 48 mg	Protein: 16.07 g	Sodium: 58 mg

HAKE (*MERLUCCIUS MERLUCCIUS*)

Dark gray to brown down the head and back with a white and silver underside. The body is long and skinny with two spiky dorsal fins, bulging eyes, and a black mouth. Found in both hemispheres, they range from 12 to 24 inches and some species can weigh up to 60 pounds. Underutilized in the market, they are line and trawler-caught and thus susceptible to net damage. The delicate flesh is soft to the touch with a white to light red tinge. As the flesh ages or is subjected to overicing or water, it sags and rapidly loses its firmness. Sold in fillets or small steaks, they are good baked, fried, pickled (escabeche), or poached. In Spain, hake are cooked and served with a variety of sauces.

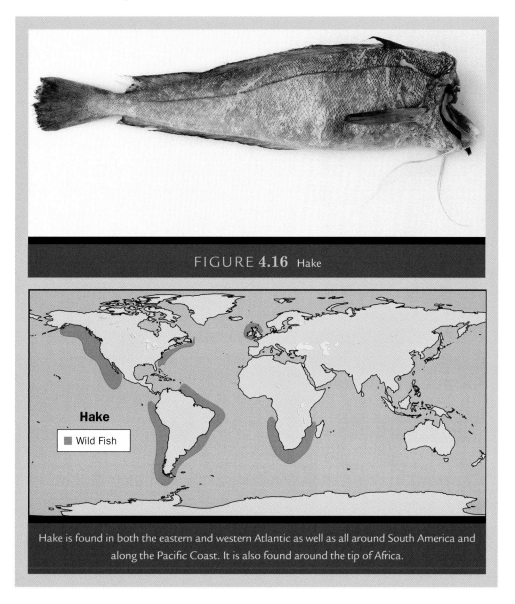

FIGURE 4.16 Hake

Hake is found in both the eastern and western Atlantic as well as all around South America and along the Pacific Coast. It is also found around the tip of Africa.

HAKE (3 OZ SERVING)

Calories: 74   Total fat: 1.87 g   Saturated fat: 0.34 g
Omega-3: 0.34 g   Protein: 13.43 g   Sodium: 71 mg

WHITING (MERLANGIUS MERLANGUS)

A species of hake, whiting have a silvery gray back with a silver belly and a black spot at the pectoral fin. Its slender and pointy body is without a barbel. The soft flesh is mild in flavor and best when eaten very fresh. They range in size from 12 to 20 inches, can be easily boned through the back, and treated much like trout. It is especially good in fish soups; the meat adds a creamy consistency when it cooks down. Breading and battering before pan frying or deep frying are other methods of working with the delicate fillets, which are also ideal for mousselines.

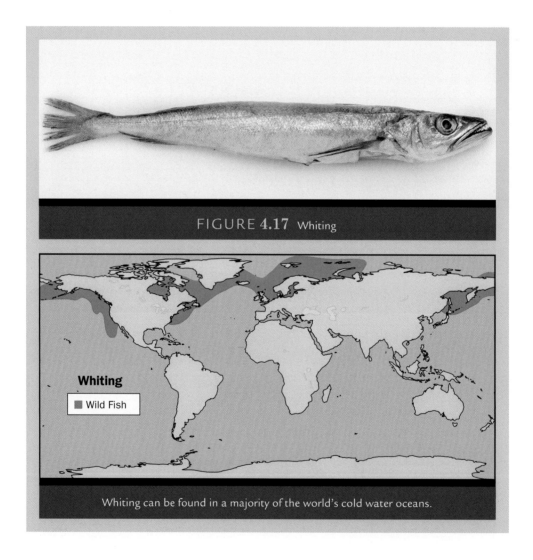

FIGURE 4.17 Whiting

Whiting can be found in a majority of the world's cold water oceans.

POLLOCK (POLLACHIUS VIRENS)

Silvery gray back with a white body and a lateral line. They have a small barbel, five fins, and grow to a length of about 3 feet. They live near the surface or on the bottom

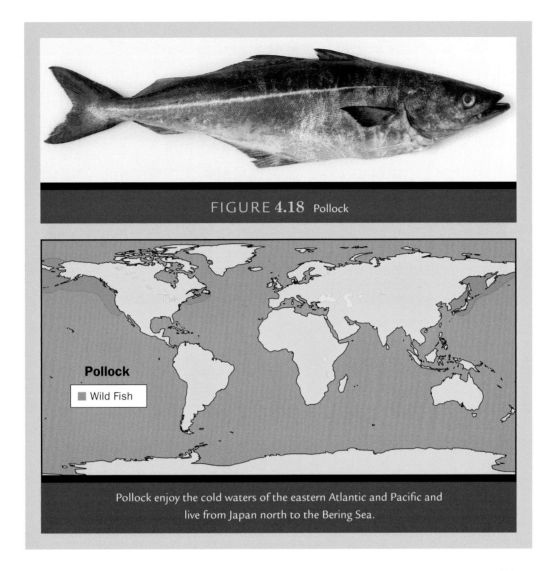

FIGURE **4.18** Pollock

Pollock enjoy the cold waters of the eastern Atlantic and Pacific and live from Japan north to the Bering Sea.

feeding on fish, crustaceans, and various marine life. Sold whole, as steaks or fillets, the meat is slightly drier than cod and is best in chowders, fried, sautéed, or stews.

POLLOCK, ALASKA (3 OZ SERVING)

Water: 69.33 g   Calories: 69   Total fat: 0.68 g   Saturated fat: 0.189 g
Omega-3: 0.461 g   Cholesterol: 60 mg   Protein: 14.60 g   Sodium: 84 mg

PACIFIC (*GADUS MACROCEPHALUS*)

Pacific cod look a lot like their eastern counterparts with a brown speckled back and white belly. Ranging in sizes up to 20 pounds, they are not as abundant or commercially important as the Atlantic cod. The flesh should be white and steaks should glisten and not appear dry. Thicker cuts from around the shoulder area are preferred for baking, broiling, grilling, pan frying, poaching, and sauté. Tail pieces are superior for chowders and deep frying.

PREPARATION METHODS

- Bake
- Broil
- Chowders
- Deep fry
- Pantry
- Poach
- Salt
- Sauté
- Steam
- Stew

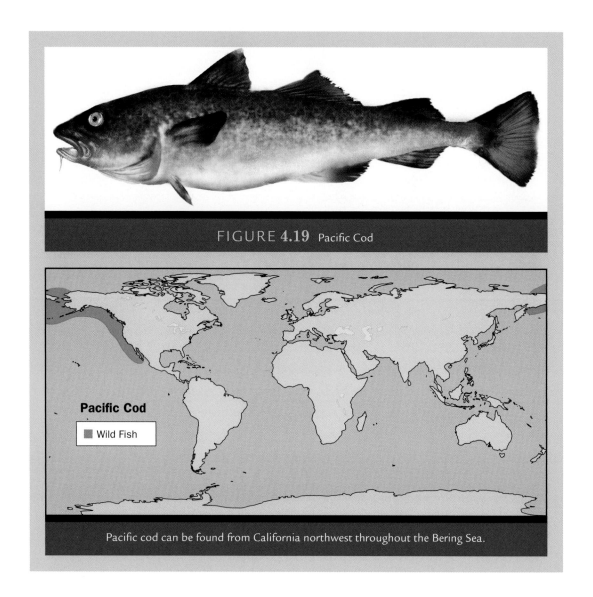

FIGURE 4.19 Pacific Cod

Pacific cod can be found from California northwest throughout the Bering Sea.

SEASONAL AVAILABILITY

Year-round

COD, PACIFIC (3 OZ SERVING)

Water: 69.09 g Calories: 70 Total fat: 0.54 g Saturated fat: 0.069 g
Omega-3: 0.235 g Cholesterol: 31 mg Protein: 15.21 g Sodium: 60 mg

DOGFISH (*SQUALUS ACANTHIAS*)

Saddled with an unappetizing name, this underutilized species has been slow to catch on with consumers. To help promote the dogfish, the Food and Drug Administration has approved the alternate name of cape shark. Ranging in size from 6 to 8 pounds, the dogfish is similar to all sharks in that its skeleton is cartilaginous not boney. It is gray to brown on top with an off-white belly. Unquestionably "shark-like" in appearance, the dogfish has five gill slits located behind the mouth. Their body shape is long and lean; they are marketed whole, headed and gutted, or fabricated into steaks or fillets. Found

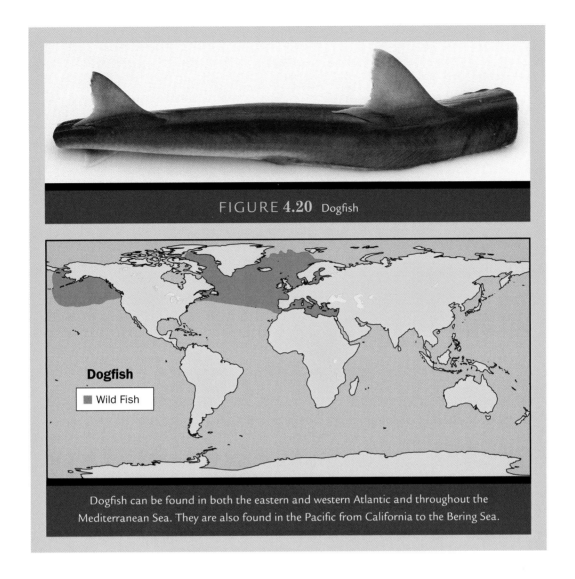

FIGURE 4.20 Dogfish

Dogfish can be found in both the eastern and western Atlantic and throughout the Mediterranean Sea. They are also found in the Pacific from California to the Bering Sea.

in the northern Pacific and Atlantic Oceans, they are caught in gill nets and by longliners. Because of their limited market, they are reasonably priced.

QUALITY CHARACTERISTICS

More accepted in Great Britain and Europe, it is commonly used to make fish and chips. Mild in flavor and fairly firm in texture, the meat is surrounded by a faint red blood line that should be removed. Like all sharks, the dogfish excretes urine through its skin and must be gutted and iced immediately upon landing to avoid a concentrated ammonia aroma and flavor. Submerging the fillets in a mixture of lemon juice and ice water will leach any off taste and aroma from the fish.

PREPARATION METHODS

- Bake
- Dried (fins)
- Sauté
- Stew
- Broil
- Fry
- Smoke

SEASONAL AVAILABILITY

Year-round

DOGFISH (3.5 OZ SERVING)

Calories: 167   Total Fat: 11.4 g   Saturated Fat: 3.8 g   Omega-3: 1.9 g
Cholesterol: 46 mg   Protein: 15.1 g   Sodium: 100 mg

Source: Seafood Handbook, their source Nettleton

EEL (ANGUILLA ROSTRATA)

Revered worldwide for their culinary versatility, eels spawn in the North Atlantic Ocean but return to fresh water where they are caught in river traps, weir pots, and nets. Resembling a snake, they average 2 to 3 feet in length and were once consumed in great numbers throughout the United States. Over the past decades, as Americans lost their taste for this interesting looking fish, much of the catch, which also includes small glass eels used for aquaculture, has been sent to Europe. Eel are found throughout the eastern Atlantic from Canada to the gulf, as well as in Asia where they are highly sought after.

QUALITY CHARACTERISTICS

Eels are very firm in texture and flavor and have a gray flesh that turns white when cooked. Because they can live several days out of water, they are often sold alive. To

FIGURE 4.21 Eel

Farmed throughout Asia, eels are commercially caught from the Carolinas and north through Canada and the Great Lakes.

CHAPTER 4 · FIN FISH IDENTIFICATION

skin an eel, first kill it and make a circular cut around the body behind the pectoral fin. Have someone hold onto the head and peel back the skin as though you were removing a wet sock. Fillet the meat off the bone or cut it into sections. Eels are high in fat so avoid rich sauces and cooking methods.

PREPARATION METHODS

- Grill
- Smoke
- Stew
- Poach
- Steam

SEASONAL AVAILABILITY

April to November

EEL (3 OZ SERVING)

Water: 58.02 g	Calories: 156	Total fat: 9.91 g	Saturated fat: 2.04 g
Omega-3: 0.161 g	Cholesterol: 107 mg	Protein: 15.67 g	Sodium: 43 mg

FLOUNDER (ORDER: PLEURONECTIFORMES)

Flounders are a boney fish that have compressed bodies with both eyes on the same side of their small head. They are bottom dwellers that have the ability to change color with their environment, and have a distinct dark and light side. Found on both coasts and throughout northern Europe and Asia, these flat fish are an important culinary species. Flounders have a market size of about 1 to 4 pounds and have many names depending on the species and region.

FLOUNDER, WITCH (*GLYPTOCEPHALUS CYNOGLOSSUS*)

The witch flounder has a small mouth, is brown in color, and can grow up to 2 feet in length.

FLOUNDER, WITCH (3 OZ SERVING)

Water: 67.20 g	Calories: 77	Total fat: 1.01 g	Saturated fat: 0.241 g
Cholesterol: 41 mg	Protein: 16.01 g	Sodium: 69 mg	

AMERICAN PLAICE (*HIPPOGLOSSOIDES PLATESSOIDES*)

American plaice have a large mouth and are red to brown in color. The tips of the anal and dorsal fins are white in color and average market size is 2 to 4 pounds.

FLOUNDER, YELLOWTAIL (*LIMANDA FERRUGINEA*)

Yellowtail flounder has a small mouth and is brown with rust-colored blotches. The fins on the non-eye side have yellow edges and they can grow up to 2 feet in length.

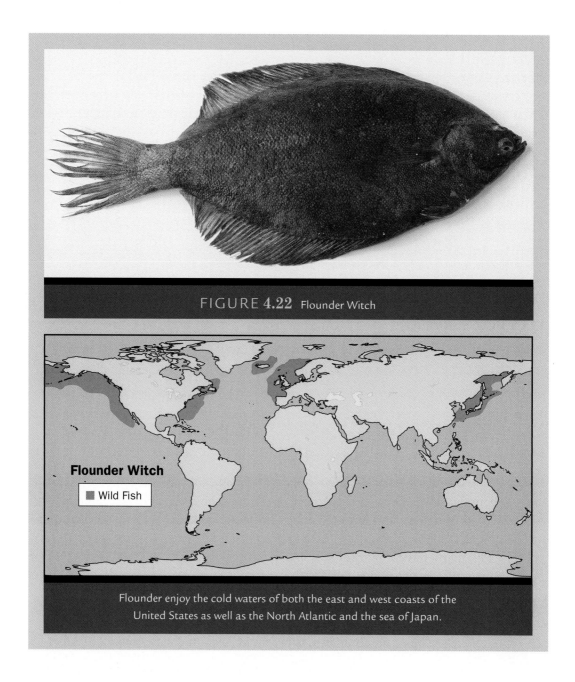

FIGURE 4.22 Flounder Witch

Flounder enjoy the cold waters of both the east and west coasts of the United States as well as the North Atlantic and the sea of Japan.

QUALITY CHARACTERISTICS FOR FLOUNDER

Flounders are delicate in both texture and flavor. Two or four fillets can be removed from the fish depending on its size or desired cooking method. Typically the fillets are skinned before cooking. Sautéing and poaching are two of the classic cooking methods for flounders; they remain moist and tender and contrast well with rich coatings or sauces, but keeping it simple is usually the best approach.

PREPARATION METHODS

- Bake
- Deep fry
- Poach
- Broil
- Pan fry
- Sauté

SEASONAL AVAILABILITY

Year-round

GROUPER, RED (*EPINEPHELUS MORIO*)

Found throughout Florida and the Caribbean, the grouper is in the sea bass family. It can grow to be several hundred pounds, but market size is between 5 and 20 pounds, depending on the species. Red grouper is most common and is dark red to brown in color, followed by the black grouper, which is usually light brown. Both species have colored blotches along their compact and strong bodies. Due to overfishing, usually with hook and line, and habitat destruction, the groupers are becoming scarce in the market place. Like other warm water species, groupers are susceptible to parasites, which are easily killed with proper cooking.

QUALITY CHARACTERISTICS

Grouper has a firm texture, large flake, and sweet mild flavor that contrast well with tropical flavor profiles. It is moist when cooked and the large fillets can easily be stuffed or butterflied.

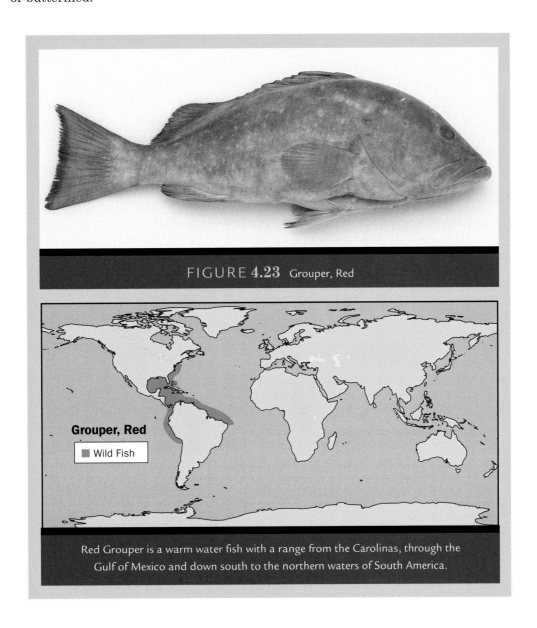

FIGURE 4.23 Grouper, Red

Grouper, Red — Wild Fish

Red Grouper is a warm water fish with a range from the Carolinas, through the Gulf of Mexico and down south to the northern waters of South America.

PREPARATION METHODS

- Baking
- Broil
- Deep fry
- Grill
- Poach
- Sauté

SEASONAL AVAILABILITY

Year-round

GROUPER, RED (3 OZ SERVING)

Water: 67.34 g Calories: 78 Total fat: 0.87 g Saturated fat: 0.198 g
Omega-3: 0.211 g Cholesterol: 31 mg Protein: 16.47 g Sodium: 45 mg

HALIBUT

- Atlantic (*Hippoglossus hippoglossus*)
- Pacific (*Hippoglossus stenolepis*)

The halibut is found in the deep cold waters of the Atlantic and Pacific. It is the largest of the flat fish and can grow to over 700 pounds but market size is much smaller. It is easily identified by its large size, and characteristic white and dark side, which can range from a mottled brown to gray. Halibut are also farm-raised.

QUALITY CHARACTERISTICS

Because of its size, the halibut can be utilized for a variety of specialty cuts. The palm-sized cheeks are nicely shaped and fairly firm. The large fillets or "fletches" are snowy white in color, are mild in flavor, and are just firm enough to hold their shape in cooking. Bone-in steaks are another common cut suitable for grilling and broiling; the bones are excellent for fish stock.

PREPARATION METHODS

- Bake
- Broil
- Fry
- Grill
- Pan fry
- Poach
- Sauté

SEASONAL AVAILABILITY

Atlantic and farm-raised halibut are available year-round.

Pacific halibut is available from March to November.

HALIBUT (3 OZ SERVING)

Water: 66.23 g Calories: 94 Total fat: 1.95 g Saturated fat: 0.276 g
Omega-3: 0.395 g Cholesterol: 27 mg Protein: 17.69 g Sodium: 46 mg

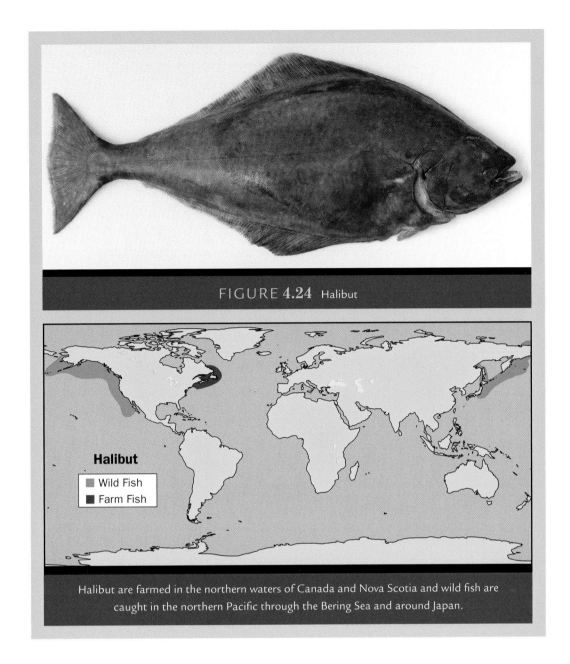

FIGURE 4.24 Halibut

Halibut are farmed in the northern waters of Canada and Nova Scotia and wild fish are caught in the northern Pacific through the Bering Sea and around Japan.

HERRING (*CLUPEA HARENGUS*)

Inhabiting the Atlantic and Pacific Oceans as well as the Mediterranean and Baltic Seas, the herring was originally salted or pickled in large quantities for human consumption, and also has commercial value as fish meal and oil. Caught with mid water trawlers, purse seine, and gill nets; they also find their way into lobster traps because they are the primary bait used by Maine lobster boats. The interesting textured roe, called *kazunoko,* is given as a New Year gift in Japan.

QUALITY CHARACTERISTICS

Although they reach a size of 1-1/2 pounds, average market size depends on their use. Small herring of several inches are canned and labeled as sardines or called *rollmops* when rolled around an onion and pickled in vinegar. The small pin bones cook down, making them soft and digestible. Fresh herring are limited to ethnic markets and big cities and vary in flavor based on their size. They should be shiny silvery blue with loose scales and a fresh briny smell.

PREPARATION METHODS

- Bake
- Broil
- Can
- Grill
- Pickle
- Salt
- Smoke

SEASONAL AVAILABILITY

June to September

HERRING (3 OZ SERVING)

Water: 61.24 g
Omega-3: 1.712 g
Calories: 134
Cholesterol: 51 mg
Total fat: 7.68 g
Protein: 15.27 g
Saturated fat: 1.734 g
Sodium: 76 mg

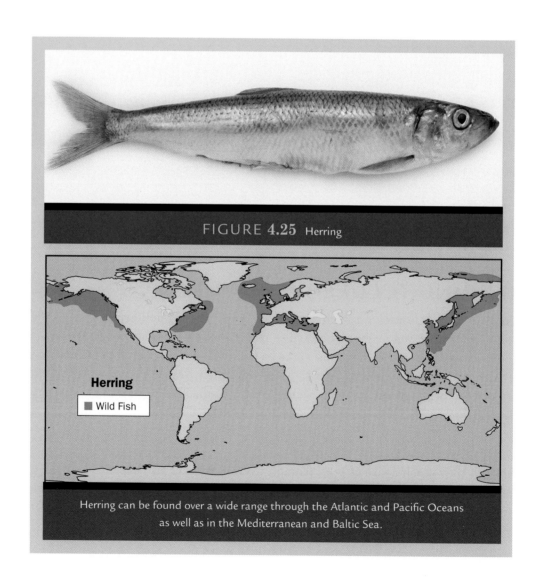

FIGURE 4.25 Herring

Herring can be found over a wide range through the Atlantic and Pacific Oceans as well as in the Mediterranean and Baltic Sea.

JOHN DORY (*ZENOPSIS OCELLATA*) (*ZEUS FABER*)

Species of John Dory are found along the East Coast from Nova Scotia to the Carolinas, throughout Europe and the Mediterranean, and in the Southern Hemisphere. They have a long and spiny dorsal fin, thin compact body, and a characteristic black spot behind the gills. It is this black spot that gives this fish its alternate name of St. Peter's fish, a biblical reference to St. Peter pulling the fish from the sea and finding a gold coin in its mouth. With this coin, he was able to pay the tax collectors, leaving his thumbprint on the fish forever. Average market size is between 3 and 8 pounds, but due to its large head and body shape, it has a very low meat-to-bone ratio.

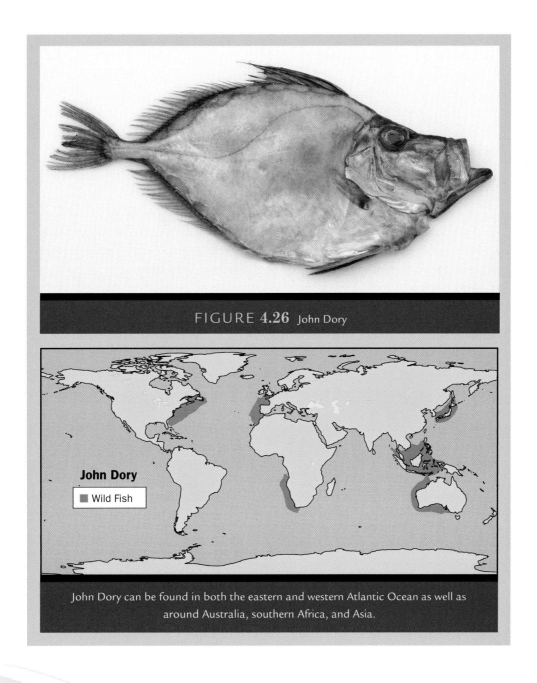

FIGURE 4.26 John Dory

John Dory can be found in both the eastern and western Atlantic Ocean as well as around Australia, southern Africa, and Asia.

QUALITY CHARACTERISTICS

John Dory is typically sold whole. It yields delicate fine-flaked white fillets similar to flounder or sole, and its bones are used for fish stocks and fumets. In the Mediterranean region, the meat is prized for Bouillabaisse.

PREPARATION METHODS

- Bake
- Grill (whole)
- Sauté
- Broil
- Poach
- Stew

SEASONAL AVAILABILITY

Year-round

JOHN DORY (3 OZ SERVING)

Calories: 81   Total fat: 0.68 g   Protein: 17.51 g

LINGCOD (*OPHIODON ELONGATUS*)

Lingcod is not a cod, but rather a greenling that inhabits the Pacific from Mexico to Alaska. Resembling cod only slightly in its shape, the lingcod can be identified by its elongated form, its long notched dorsal fin, and its large pectoral fin. It has a single lateral line and ranges in color from shades of grayish green to brown with mottled spots and a large mouth. Market size has decreased over the years, and most fish range between 10 and 20 pounds. They are caught with long lines and trawlers. Other names include green cod and buffalo cod.

QUALITY CHARACTERISTICS

The meat is light algae green when raw and turns white when cooked. Lingcod is fairly firm and dense but very mild in flavor and lean. It is perfect for those who like a mild fish.

PREPARATION METHODS

- Bake
- Fry
- Sauté
- Broil
- Poach

SEASONAL AVAILABILITY

Year-round

LINGCOD (3 OZ SERVING)

Calories: 73   Total fat: 0.76 g   Protein: 15.47 g   Sodium: 43 mg
Cholesterol: 52 mg   Saturated fat: 0.2 g

FIGURE 4.27 Lingcod

Lingcod is a Pacific fish inhabiting cold waters from California to Alaska.

MACKEREL

- Atlantic (*Scomber scombrus*)
- Spanish (*Scomberomorous maculatur*)

One of many fish in the Scombridae family, the Atlantic mackerel can be found in the western Atlantic from North Carolina to Labrador and in the east from the Baltic Sea south through the Mediterranean and into North Africa. Historically, the mackerel was a prime curing fish because of its oily meat and smaller size. It was a source of protein for early New Englanders and is still eaten salted throughout the Caribbean islands.

Easily identified by its silvery blue-gray skin, it has vertical dark lines running down the back to the lateral line, and lightening around the belly. The fish is long and slender and has small finlets similar to tuna, down their back from the dorsal fin to the tail. Market size is usually 2 to 3 pounds but smaller fish are available seasonally.

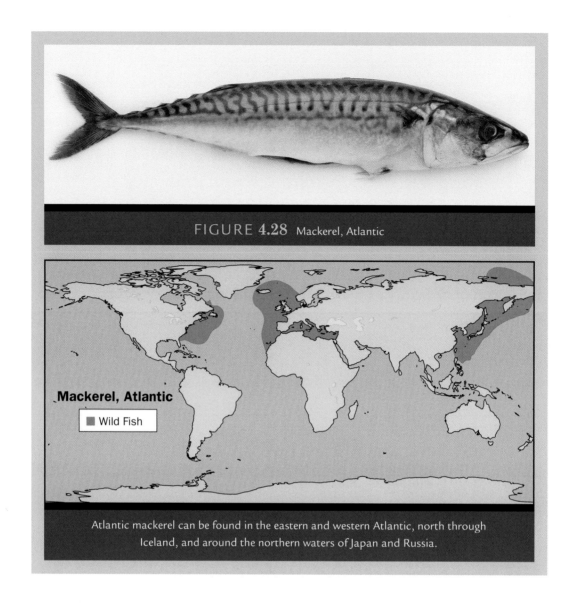

FIGURE 4.28 Mackerel, Atlantic

Atlantic mackerel can be found in the eastern and western Atlantic, north through Iceland, and around the northern waters of Japan and Russia.

Another important culinary species is the Spanish mackerel, which has a lean, lighter meat that is not as "fishy" as the darker flesh Atlantic. With weights of more than 20 pounds, it does not have the distinctive vertical lines but may have yellow-colored spots.

QUALITY CHARACTERISTICS

Mackerel are schooling fish and are commonly fished using the purse seine method. Because their body temperature can rise so much in capture, it is very important to remove the viscera and ice the fish down as soon as possible. If the fish is not chilled correctly, it can lead to scombroid poisoning. Flavor depends on feeding and spawning cycles but is considered to be full-flavored and moist with medium flake. Fillets can be skinned slightly above the blood line to remove the darkest of the meat, and marination and specific cooking techniques should be used to highlight and contrast the flavor and texture. Salted mackerel should be desalted in cold water before use.

PREPARATION METHODS

- Bake
- Broil
- Grill
- Salt
- Smoke

SEASONAL AVAILABILITY

Year-round

MACKEREL (3 OZ SERVING)

Water: 59.63 g	Calories: 134	Total fat: 6.71 g	Saturated fat: 1.910 g
Omega-3: 1.571 g	Cholesterol: 40 mg	Protein: 17.06 g	Sodium: 73 mg

MAHI MAHI (*CORYPHAENA HIPPURUS*)

The name *mahi mahi* is Hawaiian meaning "strong." This beautiful fish is also called *Dorado,* for its golden color, or *dolphin fish;* although it should not be confused with the marine mammal. Found in the tropical waters of the world, much of the commercial catch is long-lined in Southeast Asia from large factory vessels that process and freeze the fillets. Mahi are easily identified by their brilliant iridescent blue, yellow, green, and gold colors, which tend to quickly fade to a dark blue or gray when landed. They

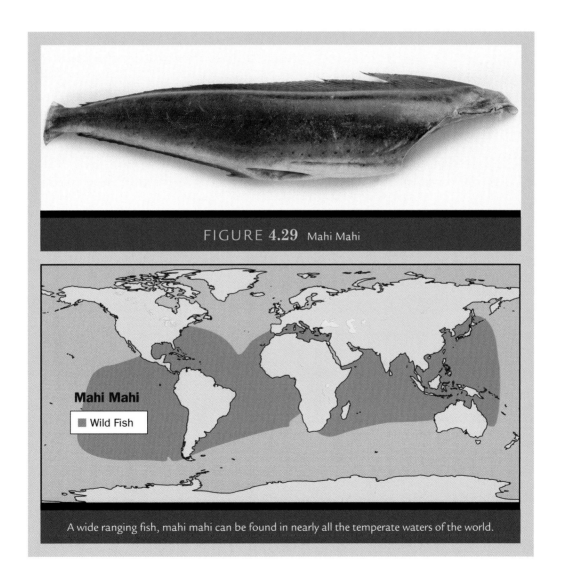

FIGURE 4.29 Mahi Mahi

A wide ranging fish, mahi mahi can be found in nearly all the temperate waters of the world.

have a blunt rounded head, a long dorsal fin, speckled sides, and a forked tail. Their average market weight is 15 to 20 pounds.

QUALITY CHARACTERISTICS

Mahi mahi have two long tapered fillets with a thick skin that can easily be pulled off by hand. The meat is gray and red in color when raw but is off-white when cooked. Each fillet contains a dark blood line, which should be removed before serving. The meat is mild but with a distinctive flavor. It is firm in texture, but not oily, making it perfect for grilling and broiling, and also it holds up to moist heat cooking. Its texture lends itself to freezing and its reasonable price makes it a bargain.

PREPARATION METHODS

- Bake
- Broil
- Grill
- Sauté
- Stir fry
- Stew

SEASONAL AVAILABILITY

Year-round

MAHI MAHI (3 OZ SERVING)

Calories: 76
Cholesterol: 73 mg
Total fat: 0.77 g
Protein: 16.07 g
Saturated fat: 0.25 g
Sodium: 109 mg

MONKFISH (*LOPHIUS AMERICANUS*)

Unique because of its long, tapering cylindrical fillet, monkfish have been referred to as poor man's lobster due to the shape and texture of the meat. Also referred to as goosefish or anglerfish, it is a perfectly engineered bottom dweller with a very large, flat wide head and large mouth with sharp piercing teeth. On top of the head dangles an unusually shaped anterior dorsal fin-like projection that can be maneuvered in front of its mouth to lure its prey, which is devoured whole or crushed with its massive mouth and jaws.

Found throughout the world but fished commercially on both the sides of the Atlantic, monkfish are sold head-off or as fillets. Its tail sections range in size from around 1 pound to more than 4 pounds. Because the heads are so large and unsightly, they rarely make it to market whole. Monkfish livers are eaten in Japan in a traditional dish called *ankimo* or in miso soup. The fillets are surrounded by a thin grayish membrane that must be removed prior to cooking, as should the narrow blood line. The meat is white and has a delicate flavor and moderately firm texture. This bottom dweller that eats all marine life has a slight shellfish flavor and is surprisingly firm for such a lean fish. Although it was once thrown back by most fishermen, today they are taken in great numbers by trawling and gillnets and are in danger of being overfished.

Because the tail meat is so firm and has a cylindrical shape similar to tenderloin of pork, they share a lot of the same applications. They can be wrapped in bacon and roasted, studded with garlic and herbs, or butterflied and stuffed. The tail meat is unique in that it does not easily flake, which makes it ideal for use in French bouillabaisse.

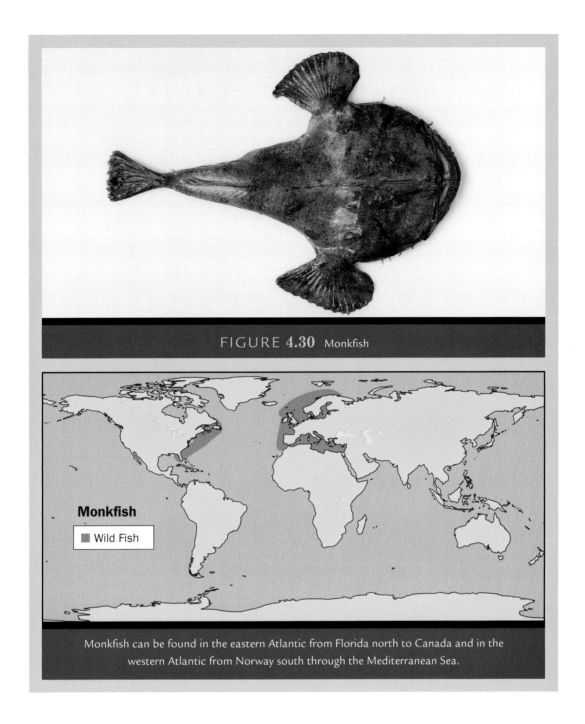

FIGURE 4.30 Monkfish

Monkfish can be found in the eastern Atlantic from Florida north to Canada and in the western Atlantic from Norway south through the Mediterranean Sea.

PREPARATION METHODS

- Bake
- Broil
- Pan fry
- Deep fry
- Sauté
- Poach
- Steam
- Smoke
- Stew

SEASONAL AVAILABILITY

Available year-round with concentrations from May to June.

MONKFISH (3.5 OZ SERVING)

Water: 70.75 g   Calories: 65   Total fat: 1.29 g   Saturated fat: 0.289 g
Cholesterol: 21 mg   Protein: 12.31 g   Sodium: 15 mg

MULLET, RED (MUGIL CEPHALUS)

There are many species of mullet that inhabit the eastern Atlantic from the Carolinas to South America as well as the West Coast. Additionally, there are many important species throughout the Mediterranean including the red mullet or rouget (*Mullus surmuletus*), which is typically grilled or used in bouillabaisse. This Mediterranean species is much leaner than the domestic striped mullet, which has a long cylindrical body with a forked tail and blunt head. It is popular throughout the southern states and is also prized for its roe.

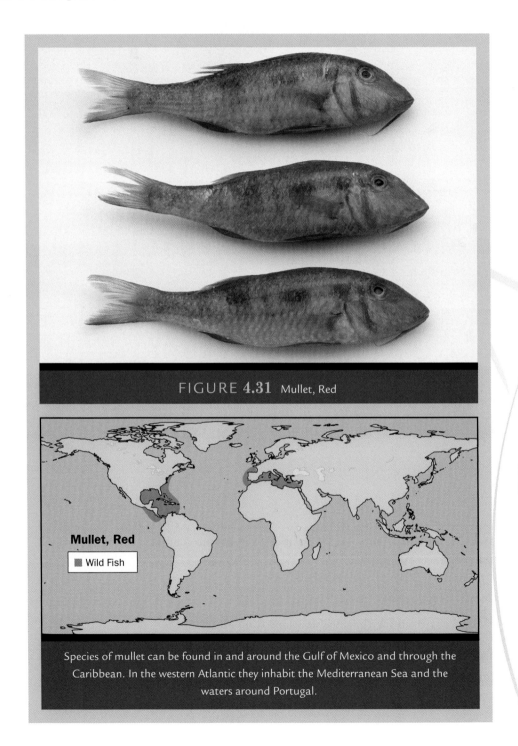

FIGURE 4.31 Mullet, Red

Mullet, Red
- Wild Fish

Species of mullet can be found in and around the Gulf of Mexico and through the Caribbean. In the western Atlantic they inhabit the Mediterranean Sea and the waters around Portugal.

QUALITY CHARACTERISTICS

The meat of the striped mullet is full flavored with a firm texture. It is oily and contains a center blood line that should be removed. Because of its high fat content, it has a limited shelf life even when frozen, so seek out fresh well-iced fish.

PREPARATION METHODS

- Broil
- Grill
- Smoke
- Deep fry
- Sauté

SEASONAL AVAILABILITY

Available year-round with concentrations from October to January.

MULLET (3 OZ SERVING)

Water: 65.46 g   Calories: 99   Total fat: 3.22 g   Saturated fat: 0.949 g
Omega-3: 0.279 g   Cholesterol: 42 mg   Protein: 16.45 g   Sodium: 55 mg

OPAH (LAMPRIS GUTTATUS)

Also called the *Hawaiian moon fish,* the opah has a disc-like compressed body shape with a beautiful gunmetal blue coloration on its back and red to orange belly. The fish has a deep keel with a proportionally short tail. White spots cover the body and the fins are a bright red.

Opahs are not schooling fish so harvesting is mainly from longline fishing. They are thought to live in the open ocean at various depths, occasionally congregating with tuna for protection; they can reach over 600 pounds. Due to its solitary nature, not much is known about its sustainability. Commonly found throughout the Hawaiian islands, the opah inhabits tropical and temperate waters of most oceans.

QUALITY CHARACTERISTICS

Originally a species that had no market value, it is now prized for its diversity of flavor and texture and has replaced other overfished Hawaiian species on restaurant menus. Because of its body shape and means of propulsion, the opah has a variety of flesh types. Beginning behind the head and running along the back is a yellowish tenderloin. Along the belly and breast area are two fibrous sections that are pink to red in color, and in larger fish the cheeks are a dark kidney red. These dark areas cook up to a brown color, whereas the loin and more tender parts are white when cooked. Overall the flesh is very firm and buttery in texture with a large flake and distinct but not overwhelming flavor, not as pronounced as that of tuna. The tender upper loin section is used for sashimi or cooked medium rare. Opah is marketed whole or headed and gutted in the 20 to 50 pound range depending on the source. It is also sold fresh as fillets or loins, which have a relatively long shelf life; it is seldom available frozen.

PREPARATION METHODS

- Bake
- Fry
- Smoke
- Broil
- Sauté
- Steam

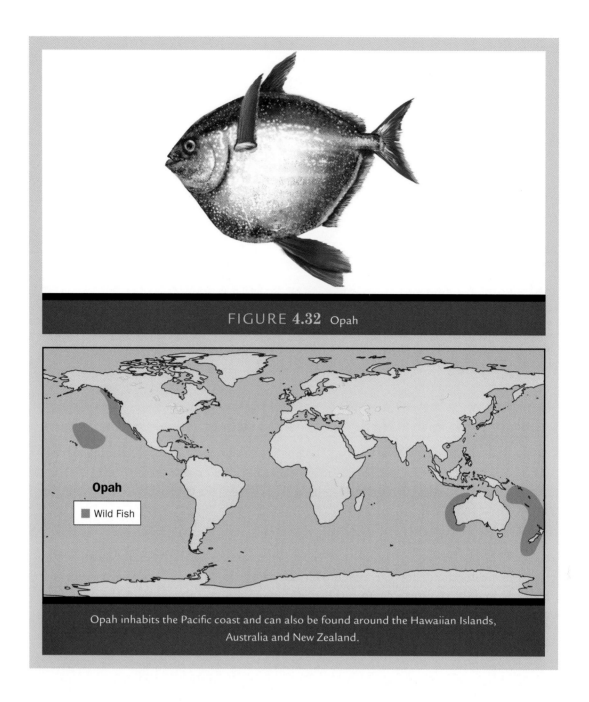

FIGURE 4.32 Opah

Opah inhabits the Pacific coast and can also be found around the Hawaiian Islands, Australia and New Zealand.

SEASONAL AVAILABILITY

Available year-round with concentrations from April to August.

OPAH (3.5 OZ SERVING)

Calories: 112   Total fat: 1.9 g
Omega-3: 0.4 g   Protein: 23.6 g

ORANGE ROUGHY (*HOPLOSTETHUS ATLANTICUS*)

The orange roughy is a beautifully colored fish with rough scales and pointed fins that lives in the deep cold oceans of the world. New Zealand fishermen were the first to drop their bottom trawling nets to depths of over 500 feet to harvest this slow-growing fish with a late reproductive age. Its head structure is tight and boney, and its noticeable bright orange/red color fades and yellows after landing. Originally called the slime fish, its name was changed for marketing reasons.

It is thought that the average market size fish of 3 to 4 pounds is roughly 25 to 50 years old and that over one million pounds of the fish has been harvested. Stocks have been slow to re-establish themselves after being exploited over the past 30 years. Due to its age, higher than normal mercury levels are possible.

QUALITY CHARACTERISTICS

Mild in flavor and firm in texture, fillets average 4 to 8 ounces and are very clean and white in color, making them very appealing to a large consumer base. It is important to skin fillets well and completely remove the blood line to avoid an unpleasant fatty flavor.

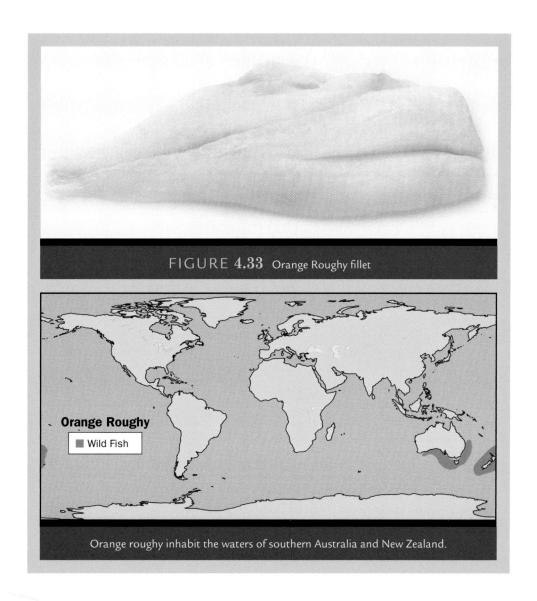

FIGURE 4.33 Orange Roughy fillet

Orange roughy inhabit the waters of southern Australia and New Zealand.

PREPARATION METHODS

- Bake
- Fry
- Poach
- Steam
- Broil
- Grill
- Sauté

SEASONAL AVAILABILITY

Year-round

ORANGE ROUGHY (3 OZ SERVING)

Water: 64.32 g Calories: 65 Total fat: 0.59 g Saturated fat: 0.013 g
Omega-3: 0.001 g Cholesterol: 51 mg Protein: 13.95 g Sodium: 61 mg

Source: Seafood Handbook, their source Frimodt

OCEAN PERCH, ATLANTIC (*SEBASTES MARINUS*)

Perch is a common name for a variety of fish species found in both fresh and salt water worldwide. In the Pacific Ocean there are many varieties but the Atlantic claims only one, which is actually a rock fish.

Also known as red fish, the ocean perch is a beautiful orange-red schooling fish that is caught by trawlers. Because it is brought up quickly from deep waters, the eyes are often bulging which does not affect the quality. Aside from its vivid color, it can also be identified by its large eyes and a protrusion on its lower lip. With weights of up to 3 pounds, the common market size is about 2 pounds. Perch grow and mature slowly, reaching sexual maturity at around 8 years old, and are in danger of being overfished.

QUALITY CHARACTERISTICS

Known for its very mild flavor and small off-white fillets, it is often individually or block frozen and available skin-on or skin-off. Skin-on perch should not be confused with red snapper, which has darker flesh, or red fish from the Gulf of Mexico.

PREPARATION METHODS

- Bake
- Fry
- Sauté
- Broil
- Poach
- Steam

SEASONAL AVAILABILITY

Year-round

PERCH, OCEAN (3 OZ SERVING)

Water: 66.89 g Calories: 80 Total fat: 1.39 g Saturated fat: 0.207 g
Omega-3: 0.318 g Cholesterol: 36 mg Protein: 15.83 g Sodium: 64 mg

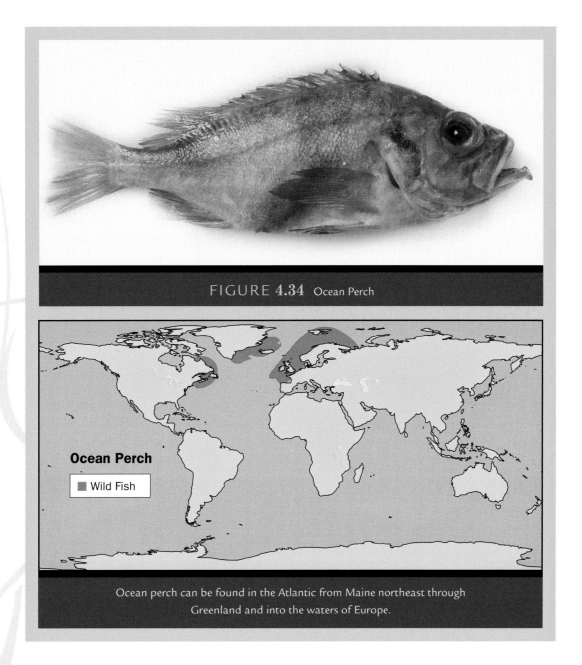

FIGURE 4.34 Ocean Perch

Ocean perch can be found in the Atlantic from Maine northeast through Greenland and into the waters of Europe.

POMPANO (*TRACHINOTUS CAROLINUS*)

Inhabiting the Atlantic coast from Virginia south through the Gulf of Mexico, this pan-sized fish ranges in size from 2 to 3 pounds and is silvery gray with a blunt nose and a noticeably forked tail. A highly-sought-after fish with a high price tag, the pompano is sometimes substituted with the inferior permit (*Trachinotus falcatus*), which will almost always be over 4 to 5 pounds.

QUALITY CHARACTERISTICS

Prized for its firm texture, small flake, and mild flavor, the pompano resembles a flat fish, but is in fact a round fish. The fillets are a light flesh color that cook up white and hold up extremely well to a variety of cooking methods. Pompano are always sold in the round.

PREPARATION METHODS
- Bake
- Broil
- Fry
- Grill
- Sauté

SEASONAL AVAILABILITY
Year-round

POMPANO (3 OZ SERVING)

Water: 60.45 g Calories: 139 Total fat: 8.05 g Saturated fat: 2.983 g
Omega-3: 0.461 g Cholesterol: 42 mg Protein: 15.71 g Sodium: 55 mg

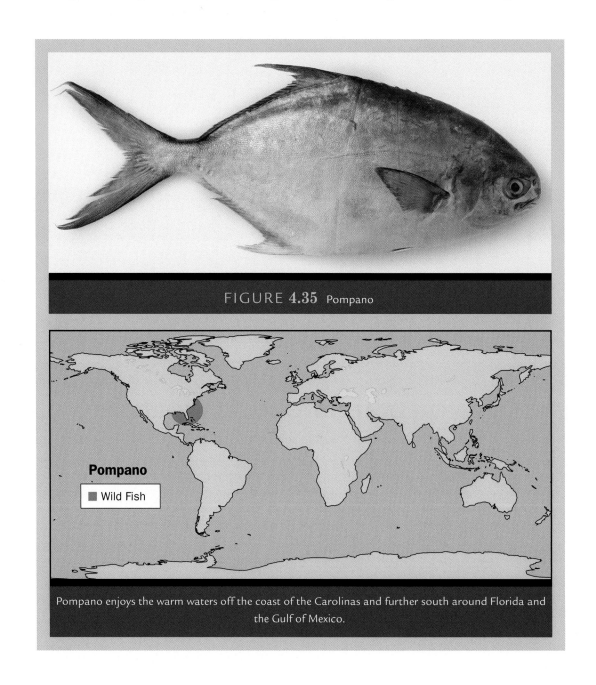

FIGURE 4.35 Pompano

Pompano enjoys the warm waters off the coast of the Carolinas and further south around Florida and the Gulf of Mexico.

PORGY (*PAGRUS PAGRUS*)

Porgies inhabit the Eastern Atlantic from Virginia south to Argentina and throughout the Mediterranean Sea. There are about fifteen species, most of which range in size from 2 to 3 pounds. An underutilized good value fish, it can be found mostly in ethnic markets.

The fish derives its name from the American Indian word meaning "fertilizer;" it has always been in important fish to the coastal populations of the Mid-Atlantic and Florida, where it is caught in abundance.

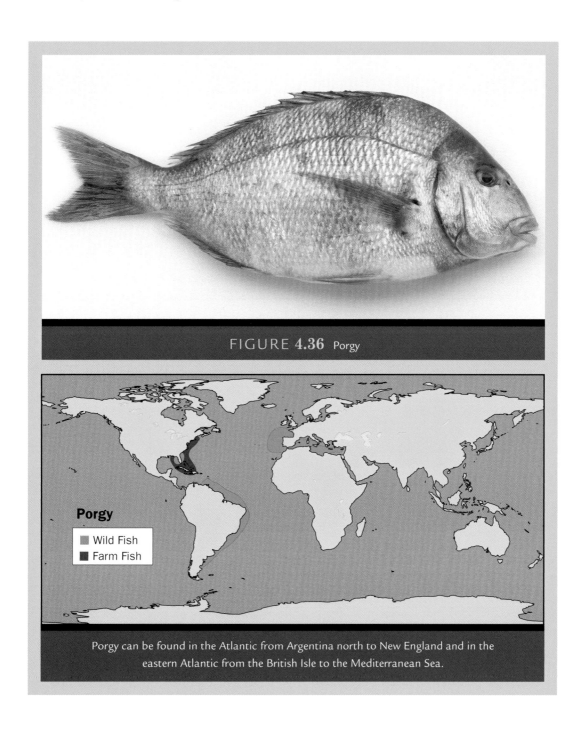

FIGURE 4.36 Porgy

Porgy can be found in the Atlantic from Argentina north to New England and in the eastern Atlantic from the British Isle to the Mediterranean Sea.

QUALITY CHARACTERISTICS

Depending on the species, porgies can range in color from silvery gray to red and have an oval-shaped body that tapers down to the tail. Also called *sea bream* or *dorade*, it is an important fish throughout the Mediterranean Sea and is often served grilled in seaside restaurants of the region.

Porgies have a mild flavor and texture similar to snapper. Because they contain a lot of bones and have a tough skin, they are typically cooked whole throughout Europe.

PREPARATION METHODS

- Bake
- Fry
- Stew
- Broil
- Sauté

SEASONAL AVAILABILITY

Year-round

PORGY (3 OZ SERVING)

Calories: 93   Total fat: 2.72 g
Protein: 15.98 g   Sodium: 53 mg

RED DRUM (*SCIAENOPS OCELLATUS*)

Commonly known as redfish or channel bass, the red drum is found from New York to Mexico with concentrations throughout the Gulf States. It can be identified by a spot on each side of the tail just behind the dorsal fin, and a lateral line along its length. It has colorations of brown to faint reddish-bronze or silvery gray on top; the belly section is white. It grows to a maximum size of about 60 pounds in the Gulf and 90 pounds in the Atlantic Ocean. Because they are adaptable to low salinity, red drum are found throughout the Louisiana coastline in marshes and lakes and are an important recreational fish in the region. Commercial fishing of red drum was halted at one time because of overfishing; it is now re-opened with strict quotas.

QUALITY CHARACTERISTICS

In the late 1970s and early 1980s, the red drum became popular throughout Louisiana as blackened redfish, a craze that extended nationally to the point it was overfished and its supply in the Gulf all but depleted.

PREPARATIONS METHODS

- Bake
- Fry
- Poach
- Broil
- Grill
- Sauté

SEASONAL AVAILABILITY

Year-round

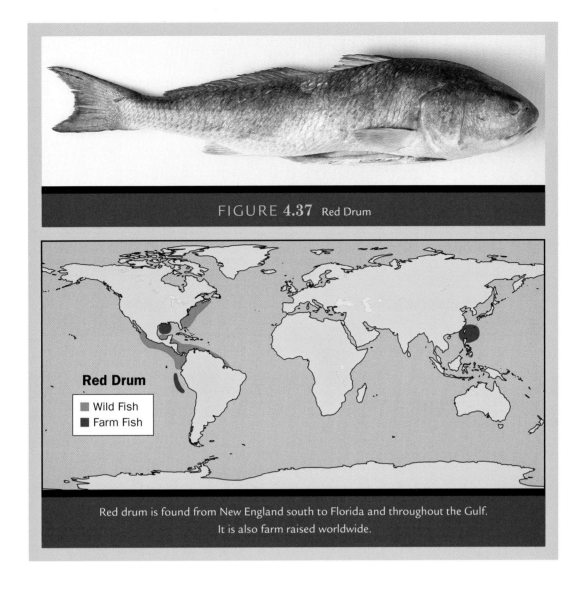

FIGURE 4.37 Red Drum

Red drum is found from New England south to Florida and throughout the Gulf. It is also farm raised worldwide.

SABLEFISH (*ANOPLOPOMA FIMBRIA*)

Also known as black cod, sablefish has grayish, mottled black velvety skin with rich buttery meat. It prefers the cold waters of the Pacific Northwest and is caught by trawlers and longliners from California to Alaska. Reaching a weight of up to 40 pounds, commercial landings tend to be in the area of 10 to 12 pounds.

QUALITY CHARACTERISTICS

Until recently, most of the catch made its way to Japan where it is highly prized for its uniquely textured and flavorful meat. Its high oil content lends itself well to smoking but also diminishes its shelf life, so a lot of the commercial catch is frozen.

PREPARATION METHODS

- Bake
- Broil
- Grill
- Sauté
- Smoke
- Steam

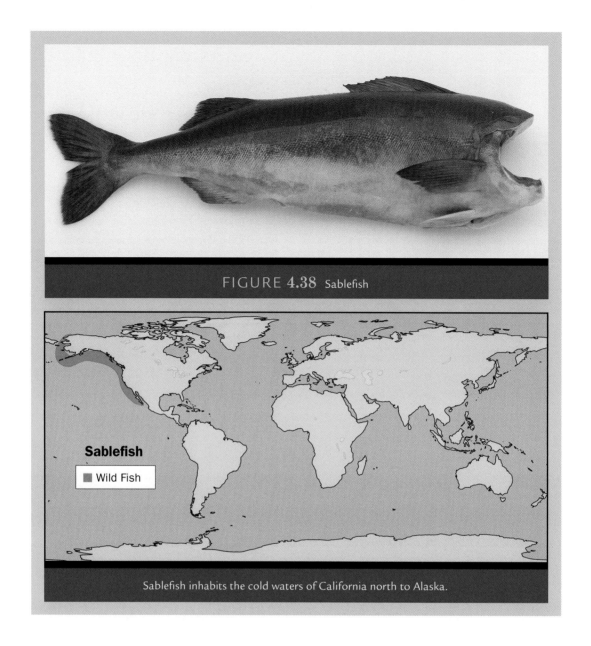

FIGURE 4.38 Sablefish

Sablefish inhabits the cold waters of California north to Alaska.

SEASONAL AVAILABILITY

Available year-round with concentrations from January to April.

SABLEFISH (3 OZ SERVING)

Water: 60.37 g Calories: 166 Total fat: 13.01 g Saturated fat: 2.721 g
Omega-3: 1.561 g Cholesterol: 42 mg Protein: 11.40 g Sodium: 48 mg

SALMON, ATLANTIC (*SALMO SALAR*)

Salmon is one of the most recognizable and popular fish in the world. It has been worshiped and prized for thousands of years for its rich flavor and fatty flesh. There are many species of Pacific salmon, but only one is native to the Atlantic. In the last century, overfishing and the continued industrialization of the coast has led the Atlantic

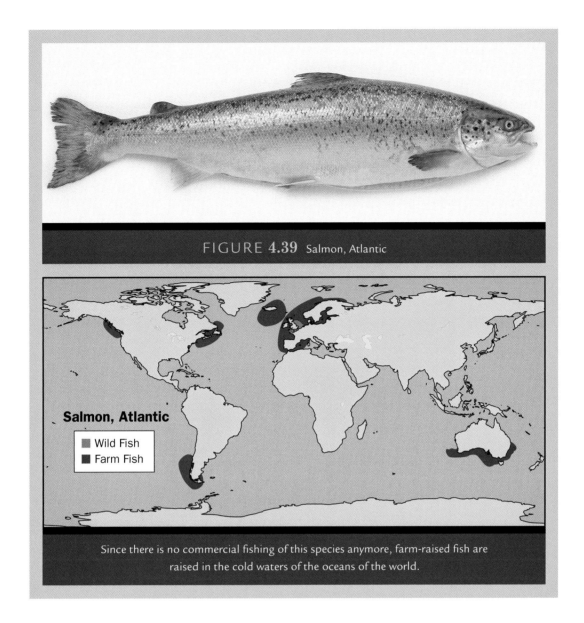

FIGURE 4.39 Salmon, Atlantic

Since there is no commercial fishing of this species anymore, farm-raised fish are raised in the cold waters of the oceans of the world.

salmon to near extinction. All salmon are anadromous, meaning they spend their lives in the open ocean, returning to fresh water rivers to spawn. Habitat destruction, including increased dams and flood barriers, increased irrigation, and polluted run-off have all contributed to the species' decline and near-extinction in the following countries: Denmark, Finland, Poland, Portugal, Spain, and the United States.

Because of this steady decline in yield, nearly 100 percent of Atlantic salmon in the market is farm-raised. It has a silver skin with small dark spots above the lateral line and a light belly.

Farm-raised Atlantic salmon can be easily identified by the underdeveloped dorsal fin and white tongue; all Pacific species have dark gray tongues. Flesh color is determined by diet and varies with all salmon. Farm-raised fish consume costly amounts of pigments such as astaxanthin in their food, which gives the meat the characteristic orange color. Atlantic fish are farmed in large floating cages throughout the world, a practice that has become controversial because of the increased algae bloom and crossover diseases between farm-raised and wild fish.

QUALITY CHARACTERISTICS

Whole fish range from 4 to 20 pounds; with 8 to 10 pounds being the most popular. Because the high oil content lends itself to curing and smoking, a wide variety of processed products are available. The flesh is firm and oily but milder in flavor than its Pacific cousin and holds up to a variety of cooking methods. Whole fish should have bright eyes and gills and smell clean. Fillets should also be firm and moist with a tight flake. Fillets that look dried out and stiff, with an open flake, may have been previously frozen. One advantage to the farm-raised Atlantic is its continuous and consistent supply. A disadvantage is its mild flavor, which is noticeably less pronounced and delicate than wild Pacific varieties.

SEASONAL AVAILABILITY

Year-round

SALMON, ATLANTIC (3 OZ SERVING)

Water: 58.57 g	Calories: 156	Total fat: 9.22 g	Saturated fat: 1.856 g
Omega-3: 1.825 g	Cholesterol: 50 mg	Protein: 16.91 g	Sodium: 50 mg

SALMON, COHO (*ONCORHYNCHUS KISUTCH*)

Coho, or silver salmon as they are sometimes called, have similar features to the Atlantic salmon, but will have a dark gray tongue and more than thirteen rays, or bones, on the anal fin; Atlantic salmon have up to ten rays on the anal fin. Cohos have silver blue skin with dark spots above the lateral line and a white belly. They are fished throughout the Pacific Northwest using gill nets, purse seines, and trolling.

QUALITY CHARACTERISTICS

The flesh of the Coho is somewhat softer in texture than other Pacific varieties and is very high in fat, lending itself to curing and smoking. Unlike sockeye, Coho are farmed in Asia and South America with average farm-raised weights between 2 and 4 pounds. Average market size for wild fish is in the 5 to 12 pound range. Flesh color of the Coho is pink to red with a mild salmon flavor and medium flake. Available throughout the summer and fall, they are marketed fresh or frozen, head-on or head-off, dressed or fabricated into fillets, steaks, or individual portions.

SEASONAL AVAILABILITY

June to October

SALMON, COHO (3.5 OZ SERVING)

Calories: 146	Total Fat: 5.9 g	Saturated Fat: 1.3 g	Omega-3: 1.3 g
Cholesterol: 45 mg	Protein: 21.6 g	Sodium: 46 mg	

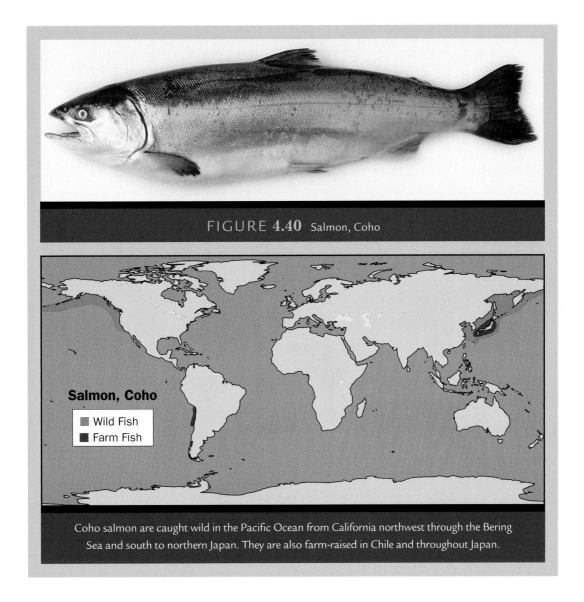

FIGURE 4.40 Salmon, Coho

Coho salmon are caught wild in the Pacific Ocean from California northwest through the Bering Sea and south to northern Japan. They are also farm-raised in Chile and throughout Japan.

SALMON, CHUM (*ONCORHYNCHUS KETA*)

Found throughout the Pacific Ocean and Bering Sea, the chum salmon can be identified by the absence of the characteristic dark spots found on all other varieties. One of the largest salmon, adults can weigh over 40 pounds and be close to 4 feet in length. Chums also have very noticeable fang-like teeth and are gunmetal blue to green with silver sides. Although chum salmon are the largest of all salmon species and are very plentiful in Alaska and Washington State, they continue to be commercially underutilized.

QUALITY CHARACTERISTICS

Chum salmon are lower in oil than most other species and the meat is firm, coarsely flaked, and rust red in color. During migration, when meat texture and skin color change, it is less desirable. Because of their wide range of quality, chums are graded from silver bright, which are ocean-caught fish, to dark chum for those fish caught in rivers. Prized for their small-sized caviar, females are processed first for their eggs and then for their meat. Chum also cures and smokes very well and, due to its lower oil content, has a longer shelf life than other varieties.

SEASONAL AVAILABILITY
August to September

SALMON, CHUM (3.5 OZ SERVING)

Calories: 120	Total fat: 3.8 g	Saturated fat: 0.8 g	Cholesterol: 74 mg
Sodium: 50 mg	Protein: 20.1 mg	Omega-3: 0.7 g	

Source: Seafood Handbook, their source USDA

SALMON, PINK (ONCORHYNCHUS GORBUSCHA)

Found in the Pacific Ocean from Asia to California, pink salmon is the most plentiful of all the species. They can be identified by their dark spots and protruding upper jaw. Flesh color is a lighter pink than many of the other salmons. It is typically canned.

SALMON, SOCKEYE (ONCORHYNCHUS NERKA) OR COPPER RIVER SALMON

Fished from the Bering Sea to California, the sockeye is considered by many to have the finest flavor and texture of any salmon. The majority are caught in gill nets in Alaska and range from 6 to 10 pounds. Sockeyes are torpedo-shaped and the mouth tapers down into a distinctive point. They are blue-green on top and are speckled rather than spotted. Sockeye salmon have had far less habitat destruction and are heavily regulated; much of the fish is sold to Japan.

QUALITY CHARACTERISTICS

Sockeye are the most-sought-after of all the salmon species and command the highest price at market. They have beautiful dark rust-colored flesh, which is high in fat and very firm in texture. The color lightens somewhat with cooking but is still a pleasing red color. High in omega-3 fatty acids, sockeye are only available from May through late September and are not farm-raised. Wild salmon may contain the small anisakis worms, which can be killed by freezing or cooking.

SEASONAL AVAILABILITY
May to late September

PREPARATION METHODS FOR ALL SALMON

- Bake
- Grill
- Raw
- Steam
- Broil
- Poach
- Smoke

SALMON, SOCKEYE (3 OZ SERVING)

Water: 59.70 g	Calories: 143	Total fat: 7.28 g	Saturated fat: 1.271 g
Omega-3: 1.046 g	Cholesterol: 53 mg	Protein: 18.11 g	Sodium: 40 mg

FIGURE **4.41** Salmon, Chum

FIGURE **4.42** Salmon, Pink

FIGURE **4.43** Salmon, Sockeye

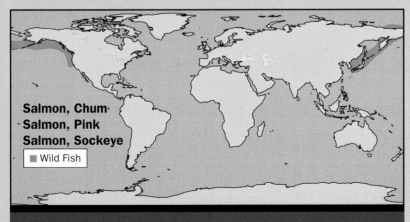

Chum salmon, pink salmon, and sockeye salmon are not farm-raised. They are caught wild from California north to the Bering Sea and south to northern Japan.

SARDINES (*SARDINELLA AURITA*) (*SARDINA PILCHARDUS*) (*HARENGULA JAGUANA*)

Sardines are a group of soft-boned fish in the herring family whose name originates from canning operations on the island of Sardinia in the Mediterranean Sea. Fished off of both coasts, a majority of the catch is centered off the cold waters of Maine and ends up in tins. Depending on the species, they range from green to black with silvery bellies. Their size varies from 3 to 10 inches.

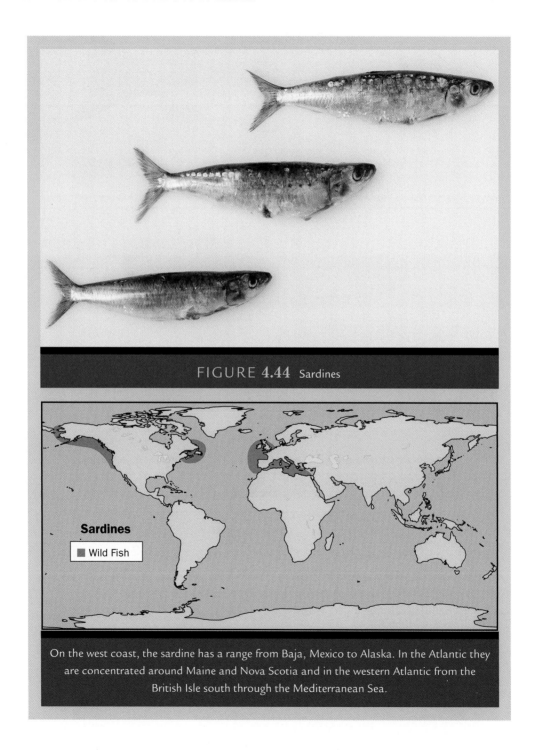

FIGURE 4.44 Sardines

On the west coast, the sardine has a range from Baja, Mexico to Alaska. In the Atlantic they are concentrated around Maine and Nova Scotia and in the western Atlantic from the British Isle south through the Mediterranean Sea.

QUALITY CHARACTERISTICS

Sardines are eaten salted, grilled, and smoked throughout Europe and the Mediterranean. In the United States, they are a reasonably priced fish but because of their size, strong flavor, and bones, are popular mainly in ethnic markets. Most are consumed canned, packed in olive oil, soybean oil, or sauces like tomato and mustard. Fresh sardines can be lightly salted or brined prior to grilling or smoking.

PREPARATION METHODS

- Bake
- Broil
- Grill
- Smoke

SEASONAL AVAILABILITY

Year-round

SARDINE (3 OZ SERVING)

Water: 50.70 g	Calories: 177	Total fat: 9.74 g	Saturated fat: 1.3 g
Omega-3: 0.835 g	Cholesterol: 121 mg	Protein: 20.94 g	Sodium: 430 mg

SKATE (*RAJA BATIS*) (*RAJA BINOCULATA*) (*GYMNURA MICRURA*)

Skate are flat body rajiformes found in seas throughout the world. Certain species are very large, but market size varieties average about 2 feet from wing to wing and weigh about 10 pounds. A delicacy in Europe, the skate was until recently considered a nuisance fish that was always thrown back by local fisherman. Skate is a bottom dweller that eats crustaceans and other marine life and is caught using a variety of fishing techniques.

QUALITY CHARACTERISTICS

The wings are the edible portion of this interesting looking species, although the liver is highly prized in Japan. Skate is one of only a few fish that are best a day or two out of the water. If cooked immediately upon landing, the meat will curl and it will be soft and spongy. The skin on the skate is thick and slimy and is typically removed after cooking, the preferred methods being poaching and sautéing. Wing meat has a faint flavor of scallops and is rather string-like but tender and juicy. At one time the meat was punched out to look like scallops, because the texture is similar, but this practice is seldom in use today. Many chefs will soak the wings in a salt or vinegar brine to remove the urea and slime.

PREPARATION METHODS

- Poach
- Sauté

SEASONAL AVAILABILITY

Year-round

SKATE (3.5 OZ SERVING)

Calories: 95　　Total Fat: 1 g

Protein: 21 g　　Sodium: 90 mg

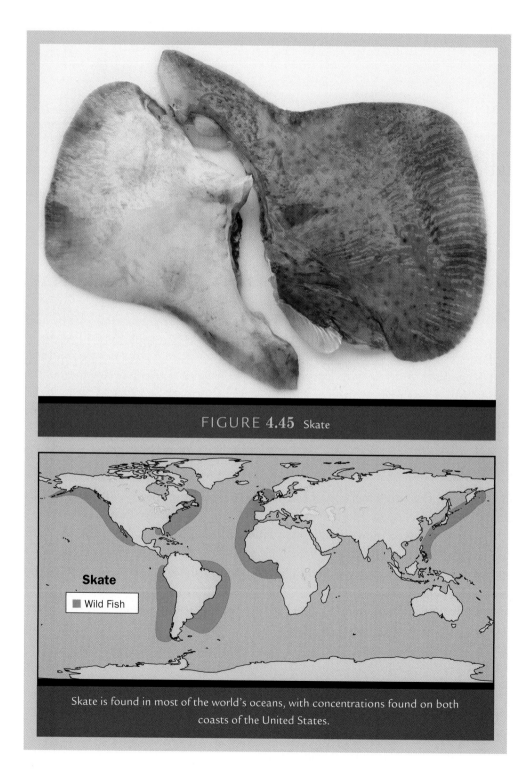

FIGURE 4.45 Skate

Skate is found in most of the world's oceans, with concentrations found on both coasts of the United States.

SMELT, RAINBOW (*OSMERUS MORDAX*)

The rainbow smelt is a long, slender translucent silvery blue fish found in both fresh and salt water from Canada to Virginia and throughout the Great Lakes. Smelt can grow to about 12 inches but market size is typically about half that. They are commercially caught with nets and are an important catch for ice fishers.

QUALITY CHARACTERISTICS

Smelts are soft-bone fish that have a very delicate fine-flaked texture and flavor. Typically eaten whole, their fresh aroma is often compared to cucumber. Because they are so delicate, they do not freeze well; whole fish should be purchased fresh. Canned smelts packed in oil or sauces are also available.

PREPARATION METHODS

- Bake
- Broil
- Fry
- Grill
- Sauté
- Smoke

SEASONAL AVAILABILITY

September to May

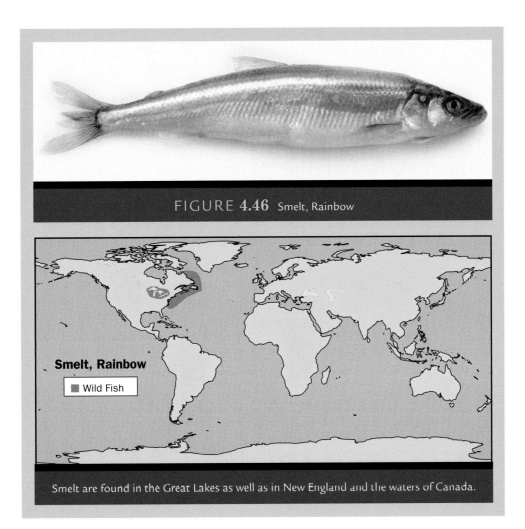

FIGURE 4.46 Smelt, Rainbow

Smelt are found in the Great Lakes as well as in New England and the waters of Canada.

SMELT (3 OZ SERVING)

Water: 66.95 g	Calories: 82	Total fat: 2.06 g	Saturated fat: 0.384 g
Omega-3: 0.756 g	Cholesterol: 60 mg	Protein: 14.99 g	Sodium: 51 mg

Source: Seafood Handbook, their source Simply Seafood

SNAPPER (GENUS: *LUTJANIDAE*)

- Black
- Blackfin
- Cubera
- Dog
- Glasseye
- Gray
- Hog
- Lane
- Mahogany
- Mutton
- Queen
- Red
- Silk
- Vermilion
- Yellowtail

Snappers are found in warm water seas throughout the world. A reef fish that also inhabits rocky bottoms, ledges, and the continental shelf, it is typically caught in waters up to 300 feet from North Carolina to Brazil, with heavy concentrations in the Gulf of Mexico. Subsisting on a diet of small fish and crustaceans, its typical market size is 2 to 4 pounds, occasionally larger. The meat is prized by chefs for its firm but delicate texture and is popular with tourists from Florida to the Caribbean. Within the family of snappers there are over 125 species, of which about 21 are found in U.S. waters. Snappers are in danger of being overfished and have been a significant by-catch of shrimp trawlers, who must now use devices in the nets to minimize snapper entrapment.

SNAPPER, RED (*LUTJANUS CAMPECHANUS*)

Of the 20 or so snappers common to the United States, it is the red snapper that is the most important. The FDA has mandated (but cannot enforce), that only true red snapper, *Lutjanus campechanus,* can be legally shipped across state lines. Because demand exceeds supply, other species are often sold as red snapper. Mislabeling is common, especially on the West Coast, where it is often labeled rock fish. Its popularity has resulted in overfishing, which is accomplished with multiple hook electric fishing gear or trawling. Gillnetting has been banned in the Gulf of Mexico where a majority of the commercial catch originates. Throughout the Caribbean islands, most are caught in fish pots or traps and to a lesser extent by spearing. Red snapper reach sexual maturity in two years and spawn from June to October. In the Gulf of Mexico, from winter through spring, the red snapper season is open for the first ten days of each month, or until the quota is reached. After this time many suppliers substitute other species for red snapper.

Careful identification of red snapper is important to ensure that you are getting what you pay for. Inferior species are routinely sold at premium red snapper prices; the subtle identifying marks are difficult to notice, especially to the ordinary customer.

A red snapper's coloring darkens with age; younger fish will be whitish-pink on the belly and light red on top. Older fish turn a unique shade between scarlet and red burgundy. All have a bright red ring around their eyes, and an anal fin that comes to a point. The triangular mouth and prominent sharp teeth contribute to its name. Noticeably lacking on the red snapper are the upper canine teeth found in dog, mutton, and mangrove varieties. Fillets are typically light brown with a visible blood line under the skin.

QUALITY CHARACTERISTICS FOR ALL SNAPPERS

Always packed in ice, fish should smell fresh and clean without any stale fishy odor. Flesh should be very firm to the touch and the eyes should be clear. Fillets should be purchased skin-on to aid in identification. The meat is mildly sweet, somewhat firm, and very moist when cooked. The color is light pink to a subtle brown, which darkens around the blood line. The color becomes noticeably whiter when cooked. Snapper is excellent baked, fried, grilled, poached, and stuffed. Its firm texture and sweet flavor combine well with coconut, scotch bonnet peppers, and various fruits such as lime, mango, and papaya.

PREPARATION METHODS

- Bake
- Fry
- Poach
- Broil
- Grill
- Sauté

SEASONAL AVAILABILITY

Available year-round with concentrations from November to May.

SNAPPER, RED (3 OZ SERVING)

Water: 65.34 g Calories: 85 Total fat: 1.14 g Saturated fat: 0.242 g
Omega-3: 0.273 g Cholesterol: 31 mg Protein: 17.43 g Sodium: 54 mg

ADDITIONAL VARIETIES

SNAPPER, GRAY (*LUTJANUS GRISEUS*), ALSO REFERRED TO AS MANGROVE SNAPPER

Gray snapper are dark with clay red on the belly, becoming grayer toward the top, with an average size of about a pound. They are abundant in the western Atlantic from the Carolinas to Brazil. The flesh is firm and very moist when cooked, with a mild but distinctively sweet taste. The fillets are pale brown with a noticeable blood line and turn white when cooked. It is a boney fish best left whole when cooking.

FIGURE 4.47 Snapper, Red

FIGURE 4.48 Snapper, Vermilion

FIGURE 4.49 Snapper, Yellowtail

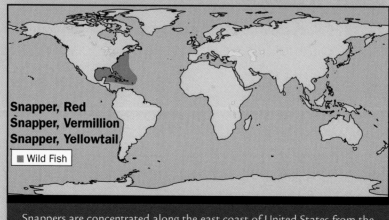

Snapper, Red
Snapper, Vermillion
Snapper, Yellowtail

Wild Fish

Snappers are concentrated along the east coast of United States from the mid-Atlantic south through the Gulf of Mexico and Carribean.

SNAPPER, MUTTON (*LUTJANUS SEBAE*), ALSO REFERRED TO AS MUTTON FISH

Mutton snapper have a white, silver, and red belly that becomes gray to silver toward the top. There is a noticeable dark lateral line and a pointed anal fin. The fish can be brightly colored depending on habitat; their average size is 4 to 10 pounds. The meat is as flavorful as the red snapper and is mild but distinctively sweet; firm and very moist when cooked. A very limited supply enters the U.S. market.

SNAPPER, VERMILION (*RHOMBOPLITES AURORUBENS*)

Also known as *beeliner,* the vermilion snapper can be distinguished from the other species by their vermilion-colored tail and eyes and a series of diagonal markings on top.

Note: The fish in Figure 4.48a has been properly gaffed on the head; avoid those fish that have gaff marks on the fillets.

SNAPPER, YELLOWTAIL (*OCYURUS CHRYSURUS*)

Yellowtail has a whitish belly and a multi colored blue-gray back with yellow spots. A bright yellow lateral line runs down the middle of the body from the mouth to the forked tail, and is an important identification marker for this species. Yellowtails are of a different genus than most snappers. The meat is slightly firm, sweet, and white; it has a smaller flake than other snappers. Depending on the Gulf Stream, they range from New England to Florida and average 2 to 3 pounds. As with all fish, proper handling and speed to market is paramount.

SOLE (FAMILY: ACHIRIDAE)

Soles can be distinguished from flounder by their small eyes, oval body, and rounded head and mouth. Many species of flounder are erroneously marketed as sole, especially because fillets are difficult to identify. In North American waters there are six species of sole, most of which are very small and have limited culinary value.

SOLE MARKETING TERMINOLOGY

SOLE, PETRALE (*EOPSETTA JORDANI*)

- Found in the Pacific from Alaska to Mexico, the petrale sole is a commercially important species on the West Coast.
- A large excellent quality Pacific flounder.

SOLE, LEMON

Common name for winter flounder caught from New England to New Jersey.

SOLE, DOVER (*SOLEA SOLEA*)

Named after the English port through which huge quantities of fish passed on their way to market, Dover sole is one of the most sought-after but expensive fish in the world. Also referred to as common or true sole, it is found in the eastern Atlantic, the Mediterranean and the North Seas, where it is usually caught with bottom trawlers. Market size range from 20 to 25 inches and should always be purchased whole. Many other varieties including flounder and Pacific sole are commonly substituted for the superior Dover sole. Whole fish are slightly slimy when fresh and an identifying dark spot can be found behind the well-developed upper pectoral fin. Color ranges from light to dark brown with mottled dark blotches. Dover sole can be distinguished from flounder by their oval bodies and small rounded, not pointed head.

QUALITY CHARACTERISTICS

In classical cuisine there is no other fish that meets the standard of Dover sole. Augusto Escoffier, in his famous French cookbook and the indispensable *Le Repertoire de la Cuisine* illustrate its importance with countless recipes and preparations. The fillets are firm and delicately sweet and buttery. For decades this one-portion fish has been sautéed table-side in the style of meunière, a delicious brown butter sauce finished with lemon and parsley. They can be stuffed or poached and served with beurre blanc. Available either whole or as single or double fillets removed from each side. The bones are used to make an excellent fish stock or fumet.

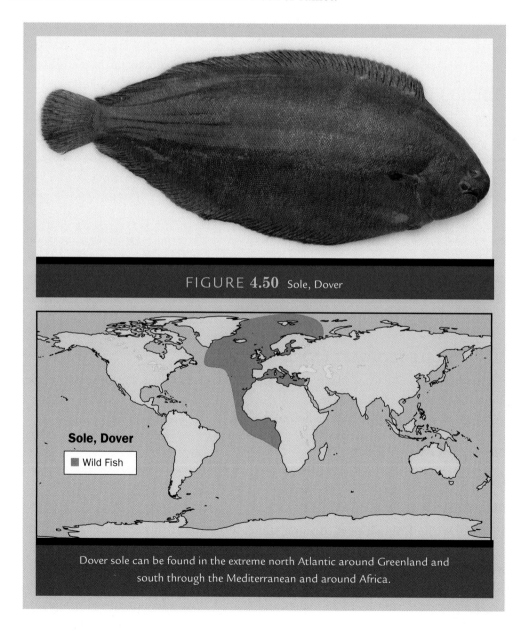

FIGURE 4.50 Sole, Dover

Dover sole can be found in the extreme north Atlantic around Greenland and south through the Mediterranean and around Africa.

PREPARATION METHODS

- Bake
- Broil
- Poach
- Sauté

SEASONAL AVAILABILITY

Year-round

SOLE (3 OZ SERVING)

Water: 67.20 g   Calories: 77   Total fat: 1.01 g   Saturated fat: 0.241 g
Omega-3: 0.085 g   Cholesterol: 41 mg   Protein: 16.01 g   Sodium: 69 mg

STURGEON

- Green (*Acipenser medirostris*)
- White (*Acipenser transmontanus*)

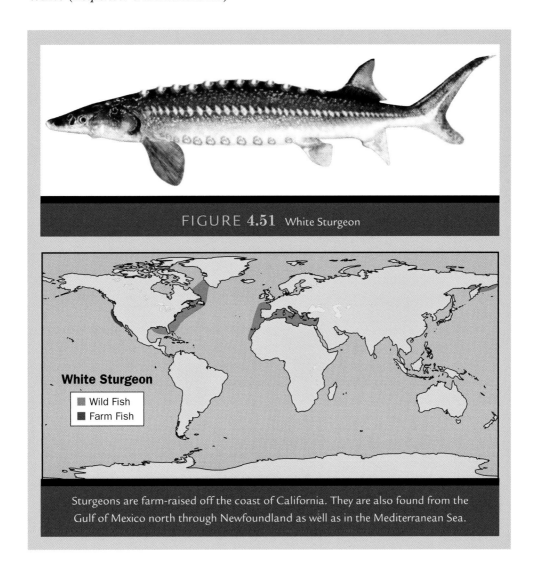

FIGURE 4.51 White Sturgeon

Sturgeons are farm-raised off the coast of California. They are also found from the Gulf of Mexico north through Newfoundland as well as in the Mediterranean Sea.

Mostly known for its roe or caviar, sturgeons were once plentiful on both coasts, but due to overfishing and habitat destruction, wild fish are very limited and most are farm-raised. These farm-raised variety range in size from 12 to 20 pounds, much smaller than their wild cousins that can grow to over 2,000 pounds and live for over 100 years. Sturgeons are cartilaginous, meaning they have no skeleton, and in place of scales, they have five rows of plates on their bodies. The head is tapered and the snout is elongated.

QUALITY CHARACTERISTICS

Green sturgeon is inferior to white and can be distinguished by rust-colored meat, green skin, and a longer snout. The white sturgeons are light gray on top and have a white belly.

Their tough, leathery skin should be removed prior to serving. The meat is extremely firm and solid: it is often compared to a light or white meat such as veal, pork, or chicken. Its flavor is very mild in comparison to its density.

PREPARATION METHODS

- Bake
- Grill
- Smoke
- Broil
- Sauté

SEASONAL AVAILABILITY

Year-round

STURGEON (3 OZ SERVING)

Water: 65.07 g	Calories: 89	Total fat: 3.43 g	Saturated fat: 0.778 g
Omega-3: 0.313 g	Cholesterol: 51 mg	Protein: 13.72 g	Sodium: 46 mg

SWORDFISH (*XIPHIAS GLADIUS*)

Because the swordfish is such a sweet-tasting, firm-fleshed fish, its prominent place on menus worldwide has caused it to be overfished. Currently, strict quotas control the catch. They are caught on longlines, gill nets, and on a smaller scale by harpooning and hand lines. Females reach reproductive age between 4 and 5 years, but as with any fish, it is difficult to target those of a specific size and many immature females are unintentionally caught in gill nets. In the late 1990s, a national campaign entitled "Give Swordfish a Break" was launched and eventually led to many chefs taking it off their menus. This awareness also led the U.S. government to ban it from being fished in over 100,000 square miles of the Atlantic Ocean. This concern for diminished supply of a specific fish resulted in widespread interest in other endangered species including Chilean sea bass. Currently, swordfish stocks are on the rebound thanks to government intervention and a ban on longlining along portions of the eastern Atlantic. High market prices also keep it in check. It is a very popular sport fish throughout Florida, Mexico, and the Caribbean.

Average market size of swordfish range from 50 to 225 pounds but they can grow to well over 1,500 pounds. Found worldwide in all but the most northern waters, they can

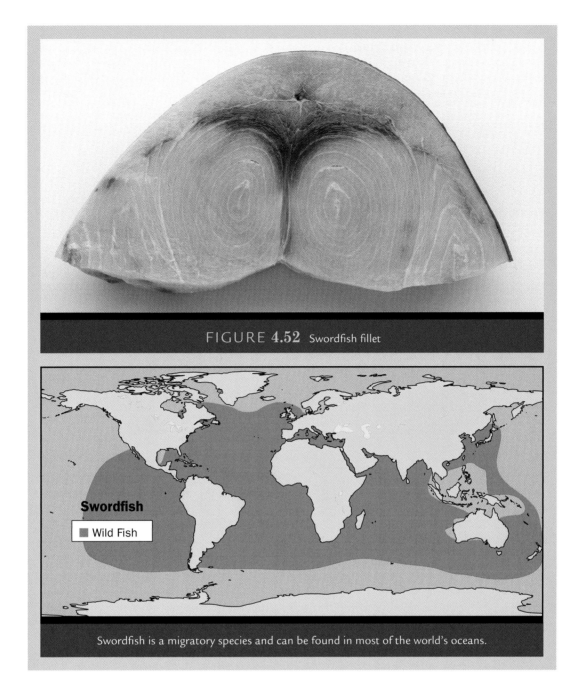

FIGURE 4.52 Swordfish fillet

Swordfish is a migratory species and can be found in most of the world's oceans.

be identified by their characteristic sword-like snout and tall rigid dorsal fin. Color is gray to black above and light brown to white below.

QUALITY CHARACTERISTICS

Swordfish feed on a variety of species including crustaceans, squid, and other fish, which contributes to their firm texture and moist flavor. Raw flesh has color variations of white, orange, and pink, and turns off-white when cooked; its thick skin is off-white to gray color. It is sold in sides (halves) or loins (quarters) as well as in steaks and wheels, both fresh and frozen. All meat should be shiny. Avoid dried-out fillets with dull or dark coloration. Because of its firm texture and high fat content, swordfish is excellent grilled, its flavor enhanced with simple ingredients such as lemon, salsas, and teriyaki sauce.

PREPARATION METHODS

- Broil
- Sauté
- Grill

SEASONAL AVAILABILITY

Available year-round with concentrations in the fall.

SWORDFISH, ATLANTIC (3 OZ SERVING)

Water: 64.28 g Calories: 103 Total fat: 3.41 g Saturated fat: 0.932 g
Omega-3: 0.696 g Cholesterol: 33 mg Protein: 16.83 g Sodium: 76 mg

TILAPIA (*TILAPIA NILOTICA*)

Tilapia is second only to carp as the world's most farm-raised fish. There are many species listed under this common name, with *Tilapia nilotica* being well suited for aquaculture. Quick to grow, they withstand a wide range of fresh and salt water conditions and are omnivorous, preferring vegetation, thus reducing the production costs. This hardy fish is raised in various size aquatic facilities throughout Asia, Africa, and Latin America. Historically, there are many biblical references to tilapia, and it is referred to as St. Peter's fish. Even today, it is a truly sustainable method of supplying impoverished areas with low-cost, high-protein, healthy food.

Shaped like a sunfish but with a slightly longer body, the fish are varying shades of mottled black, gold, red, and white. They are available live in many ethnic restaurants.

Ranging between 1 and 3 pounds for whole fish, tilapia has fillets that are very mild in flavor and can be easily overpowered by heavy sauces and seasonings. The meat is white with hues of pink that spread out from around the dark blood line between the skin and meat. Like many closed-system farm-raised fish, the flavor characteristics of tilapia can be muddy if the water quality is not monitored. Frozen product is available in a variety of packaging and fillet sizes up to 9 ounces. Because of their moderately firm texture, they freeze well and have an overall clean appearance suitable to the occasional fish eater.

PREPARATION METHODS

- Bake
- Fry
- Steam
- Broil
- Sauté

SEASONAL AVAILABILITY

Year-round

TILAPIA (3 OZ SERVING)

Water: 66.37 g Calories: 82 Total fat: 1.44 g Saturated fat: 0.651 g
Cholesterol: 42 mg Protein: 17.07 g Sodium: 44 mg

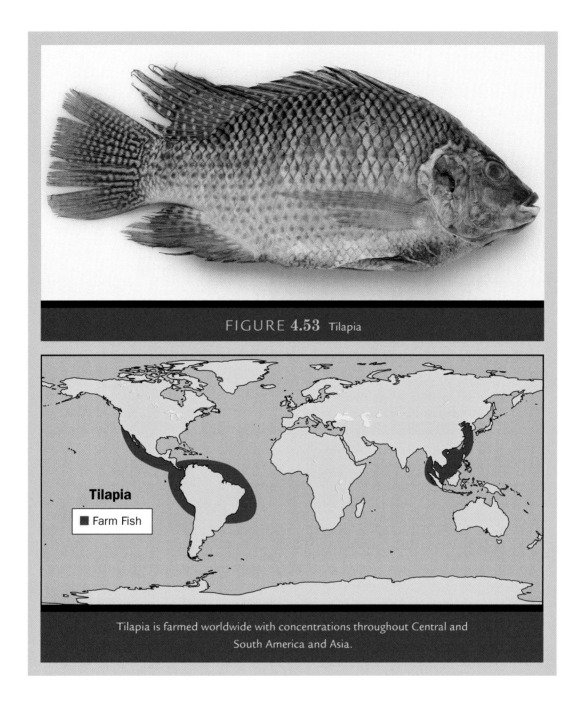

FIGURE 4.53 Tilapia

Tilapia is farmed worldwide with concentrations throughout Central and South America and Asia.

TILEFISH (*LOPHOLATILUS CHAMAELEONTICEPS*)

A beautiful fish ranging in colors from pink to green and yellow, the tilefish inhabits deep waters in the Gulf of Mexico and along the Continental Shelf from Nova Scotia to Florida. It has a blunt head and corpulent body with a noticeable fleshy protrusion in front of the dorsal fin.

QUALITY CHARACTERISTICS

Subsisting on a diet of crabs, tilefish has flesh that is firm and mild. It has a market size of 5 to 10 pounds but larger fish are common. When raw, the meat has a pinkish-brown color, which turns white when subjected to heat. The fillets are nice and thick which keeps them moist when cooked.

PREPARATION METHODS

- Bake
- Fry
- Sashimi
- Steam
- Broil
- Poach
- Sauté

SEASONAL AVAILABILITY

Year-round

TILEFISH (3 OZ SERVING)

Water: 67.07 g Calories: 82 Total fat: 1.96 g Saturated fat: 0.375 g
Cholesterol: 42 mg Protein: 14.88 g Sodium: 45 mg

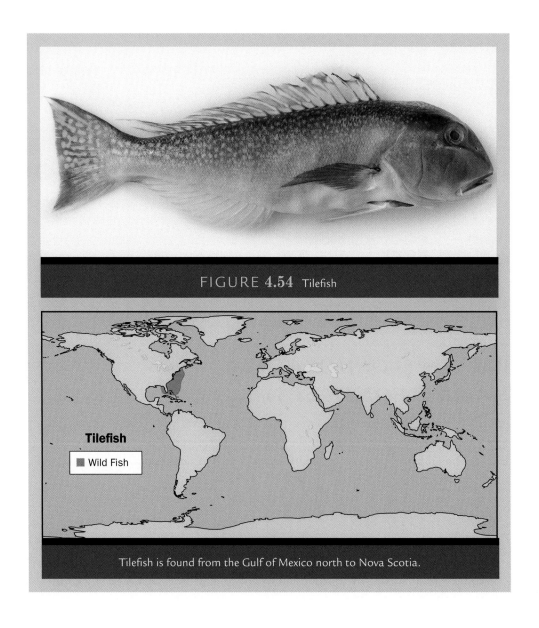

FIGURE 4.54 Tilefish

Tilefish is found from the Gulf of Mexico north to Nova Scotia.

TROUT, RAINBOW (*SALMO GAIRDNERI*)

Named for the rainbow-colored band that runs laterally from head to tail, it has an olive green to brown back with black speckled spots all along its body, head, and fins. Rainbow trout sold commercially in the United States are predominantly farm-raised in ponds and concrete raceways structures, and harvested at about 8 to 16 ounces. Because of their small size, they are commonly sold whole, dressed, or butterflied.

In the wild, trout prefer clean cold water. They are found in the western part of the United States from Mexico north to Alaska, where they are known to migrate from rivers and streams into the ocean.

QUALITY CHARACTERISTICS

The fillets can be pale white to pink depending on their diet, which is controlled through the commercial feeding process to reach market size in just under a year. Delicately

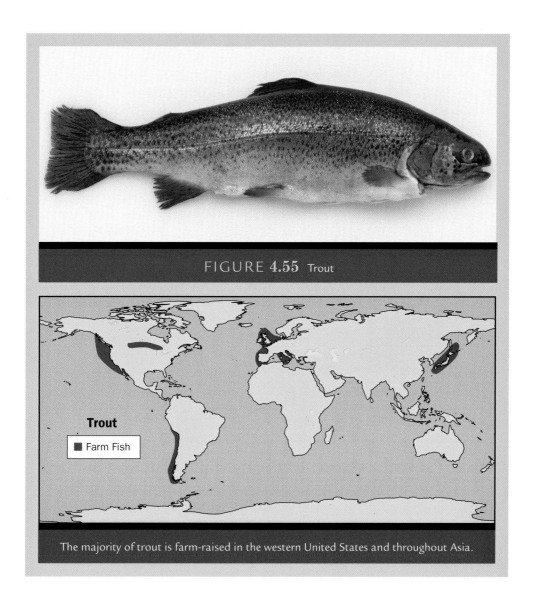

FIGURE 4.55 Trout

The majority of trout is farm-raised in the western United States and throughout Asia.

flavored and textured, the fillets are thin with a small flake. They are thought by some to have a slight earthy flavor that contrasts well with nuts, butter, herbs, and citrus.

PREPARATION METHODS

- Bake
- Fry
- Sauté
- Broil
- Poach
- Smoke

SEASONAL AVAILABILITY

Year-round

TROUT (3 OZ SERVING)

Water: 60.71 g Calories: 126 Total fat: 5.62 g Saturated fat: 0.977 g
Omega-3: 0.796 g Cholesterol: 49 mg Protein: 17.65 g Sodium: 44 mg

TUNA OVERVIEW

Most people do not realize how large and powerful tuna can be. For centuries they have wandered the globe with great speed, determination, and velocity. Their historic importance to the cultures of the Mediterranean is evident in the symbols and writings of the time. Each spring the fishermen of Favignan, Sicily, ready their nets for the annual matanza. This ancient fishing technique, which translates to "massacre," involves setting out an elaborate series of traps and fixed nets which herd the 400 pound giant bluefins from their migration path into a series of successively smaller traps until they reach the final net or "death chamber." Here they are hoisted and gaffed out of the water; quickly meeting their *coup de grace*.

Tuna inhabit the Atlantic and Northwest Pacific oceans. Accounts of their yearly visits through the Straits of Gibraltar and into the Mediterranean can be found in Greek mythology and in the writings from Carthage and the Byzantine Empire. Once these great fish left the Mediterranean, their destination was unknown. Scientists and marine biologists have begun a tagging system that has yielded very interesting results. It has been suggested that there are two distinct tuna populations co-existing off the coast of Canada and New England. These two groups join to feed, but separate and migrate to individual spawning grounds.

Prior to the modern fishing and canning industry, tuna had few detractors, its hierarchy on the food chain assured by its bullet-like design and powerful ability to engage its prey. Some tuna species are warm-blooded (endothermic) allowing them to increase their internal body temperature above the water temperature, creating a more diversified range of conditions and feeding opportunities. Resourceful as they are, tuna are no match for the modern fishing vessel. Satellite tracking and global positioning can guide these massive factory ships on an intercept course. The world's hunger for this unique meat has led to commercial depletion of several species. World fishery management groups realize the imminent threat and have set regulations and quota guidelines aimed at reducing and controlling the catch, but it may be too late. Until world agencies can devise a method of certification that chefs and consumers can understand and follow, the future of this highly prized fish remains in question.

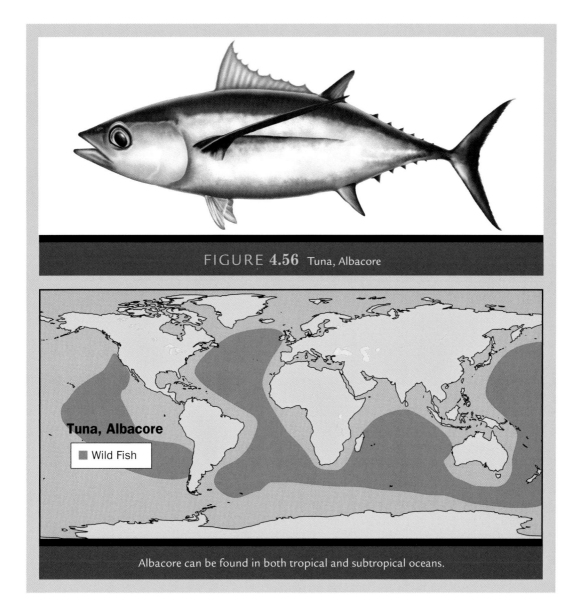

FIGURE 4.56 Tuna, Albacore

Albacore can be found in both tropical and subtropical oceans.

In addition to the ancient matanza, tuna are captured using longline, pole and line, and purse seine, a technique that surrounds the school with a net that is "pursed" or pulled up at the bottom. Once secured, this net is dragged close to the vessel and the fish are dumped or scooped onto the boat. Purse seine fishing tends to bruise the fish more than other methods. It is destructive to the other non-target fish and marine life, including sea turtles and dolphins that follow the tuna to feed and avoid sharks.

The United States is one of the largest consumers of canned tuna. Most American children are accustomed to eating tuna fish salad sandwiches. As a nation, each of us consumes about 3 pounds a year. In the 1960s and 1970s, when pods of dolphins were killed by careless fishermen, a public outcry led to a dolphin-safe movement. Currently, the National Marine Fisheries Service has developed tracking and verification regulations in the eastern tropical Pacific Ocean to ensure the accuracy of the "dolphin-safe label."

Tuna are extremely fast swimmers able to propel themselves effortlessly through the water. This maneuvering, combined with a very well-designed body shape, contributes to the wide range of color and variety of the meat throughout the tuna's body. Tuna is purchased in cuts similar to beef, with each one having a different flavor and texture. The belly portion of the fish is referred to as *toro* in Japanese, with the pinkish front portion considered the prime cut. It has an extremely rich and fatty texture and is enjoyed raw. Above the toro running along the back is the cut called *akami*. Both portions vary in grade and price throughout the length of the fish.

TUNA (FAMILY: *THUNNUS*)

Albacore:	*Thunnus alalunga*
Bigeye:	*Thunnus obesus*
Blackfin:	*Thunnus atlanticus*
Longtail:	*Thunnus tonggol*
Northern bluefin:	*Thunnus thynnus*
Pacific bluefin:	*Thunnus orientalis*
Southern bluefin:	*Thunnus maccoyii*
Yellowfin:	*Thunnus albacares*
Bonito:	*Sard sard*

SEASONAL AVAILABILITY

Most species are available year-round.

TUNA SPECIES IN OTHER GENERA (FAMILY: *SCOMBRIDAE*)

- Bullet
- Butterfly mackerel
- Dogtooth
- Frigate
- Kawakawa
- Little tunny
- Skipjack
- Slender

ALBACORE (*THUNNUS ALALUNG*)

The albacore's fame has come from its place on grocery store shelves as the premium white meat canned tuna. All canned tuna in the United States labeled white meat must come from the albacore. Found in both tropical and subtropical waters of the Atlantic and the Pacific, the average market size is between 10 and 30 pounds, reaching sexual maturity at just over 3 feet in length. It can be identified by its long pectoral fin and its yellow dorsal finlets. The belly is a solid off-white color and the back is a silvery dark blue to gray. The meat color is the lightest of all tuna and ranges from a light brown or beige to dark brown, although when cooked it is more white than brown. The meat has a large flake and is rich and beef-like and very moist when not overcooked. Typically the albacore is not the first choice for sashimi because its flesh is not as dense and firm as the bluefin, bigeye, or yellowfin.

TUNA, ALBACORE (3 OZ SERVING)

Water: 62.21 g	Calories: 109	Total fat: 2.52 g	Saturated fat: 0.673 g
Omega-3: 0.733 g	Cholesterol: 36 mg	Protein: 20.08 g	Sodium: 320 mg

BIGEYE (*THUNNUS OBESUS*)

Bigeye tuna are highly prized for their mild flavor, and rich fatty flesh that is light-to-medium burgundy color that oxidizes quickly when exposed to air, which is why they are nearly always sold fresh as loins, or canned. Also referred to as *ahi* in Hawaii and the United States, bigeyes are a deep-water tuna found in nearly all the world's waters with the exception of the Mediterranean Sea.

Often misidentified as yellowfin tuna, bigeyes are dark shiny blue/black on the back and have a gray to white underside with yellow finlets and a dark rear dorsal fin. Ranging in size from 20 to 400 pounds, they are one of the largest tuna species. They can be identified by their large head and big eyes. Because they are so close in appearance to yellowfin especially in small fish, to accurately identify them, count the gill rakers,

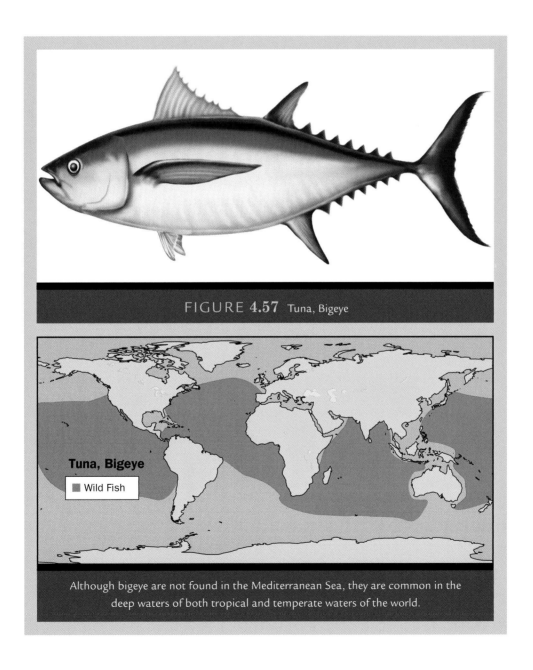

FIGURE 4.57 Tuna, Bigeye

Tuna, Bigeye
■ Wild Fish

Although bigeye are not found in the Mediterranean Sea, they are common in the deep waters of both tropical and temperate waters of the world.

TUNA OVERVIEW

which are finger-like cartilaginous projections located in front of the gill arch; there should be 18 to 22 gill rakers.

TUNA, BIGEYE (3 OZ SERVING)

Water: 60.34 g	Calories: 92	Total fat: 0.81 g	Saturated fat: 0.200 g
Omega-3: 0.237 g	Cholesterol: 38 mg	Protein: 19.87 g	Sodium: 31 mg

BLACKFIN (*THUNNUS ATLANTICUS*)

Blackfin are a small species of tuna ranging in size from 10 to 20 pounds with a maximum weight of up to 50 pounds. They are a warm water fish that have a football-shaped body with a black back, and a yellow-to-dark-gold strip running the length of the body. Their belly is white and the finlets behind the dorsal and anal fins are dark. Blackfin tuna are mild in flavor and have a light color when cooked.

LONGTAIL (*THUNNUS TONGGOL*)

Longtail range in size from 20 to 30 pounds and are a tropical species that enjoy the warmer waters of Asia and Northern Australia, where all commercial fishing of the species has been banned. Longtail can easily be identified by their long and slender tail and their short pectoral fin. A dark streak runs down the very top and the remaining body is a mix of silver and white.

ATLANTIC BLUEFIN (*THUNNUS THYNNUS*)

Inhabiting both sides of the Atlantic, and into the Mediterranean Sea, the bluefin is the largest of the scombridae species with adults ranging from 400 to 1,500 pounds. The fish is a blue gunmetal color on top with a silvery gray white belly and a large mouth. They are highly migratory fish known to spawn in the Gulf of Mexico and the Mediterranean Sea.

Bluefin have been a high value fish only since the 1970s when Japanese consumption increased; prior to that, it was sold as cat food. Currently the bluefin is being overfished and future supplies are in question. Both the United States and the European Union are working on a bluefin tuna recovery program to establish quotas and end overfishing. Many factors contribute to the great variety of flavors and textures found throughout the fish. The large fish is fabricated into many sections or cuts, many of which are unique and diverse. Because of its size, the flavor and texture of the meat is distinctly different. The well-worked muscles surrounding the fins taste different than the four distinctive lateral loins or the meat scraped from the bones. In the tskuji market in Tokyo, the tails are cut and the meat is examined for its color and fat content. Expert fabrication with large sword-like knives separates the large tuna into precise quarters. The loins are further fabricated and sold immediately in sushi restaurants and grocery stores throughout Japan.

SOUTHERN BLUEFIN (*THUNNUS MACCOYII*)

Southern bluefin tuna are found in cold and temperate waters. Adults can reach 500 pounds. The back is dark and silvery with a series of downward stripes and a silvery white belly with yellow finlets with a dark border. Like its northern cousin, the southern bluefin is in danger of being overfished.

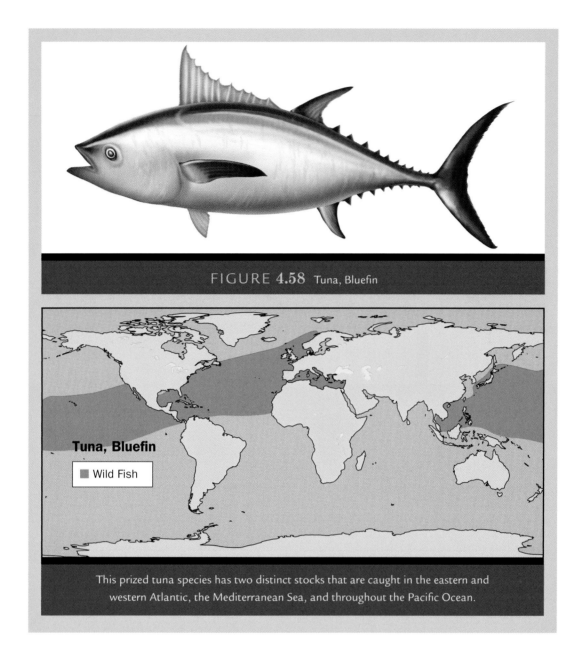

FIGURE 4.58 Tuna, Bluefin

This prized tuna species has two distinct stocks that are caught in the eastern and western Atlantic, the Mediterranean Sea, and throughout the Pacific Ocean.

TUNA, BLUEFIN (3.5 OZ SERVING)

Calories: 144   Total fat: 4.9 g   Saturated fat: 1.3 g   Omega-3: 1.3 g
Cholesterol: 38 mg   Protein: 23.3 mg   Sodium: 39 mg

Source, Seafood Handbook, their source USDA

YELLOWFIN (*THUNNUS ALBACARES*)

Inhabits tropical and subtropical waters throughout the world but not the Mediterranean Sea. With adult weights of up to 400 pounds, the yellowfin can be identified by yellow anal and dorsal fins, as well as pectoral fins that are unusually long. The back is gunmetal blue, often with a yellow horizontal strip below. The belly is a silvery blue color, crossed by vertical lines. It is marketed as ahi in Hawaii and the United States where commercial landings are monitored. Although this species reproduces quickly, it is still in danger of being overfished. With bluefin stocks declining, it is only a matter of time before this beautiful fish is threatened.

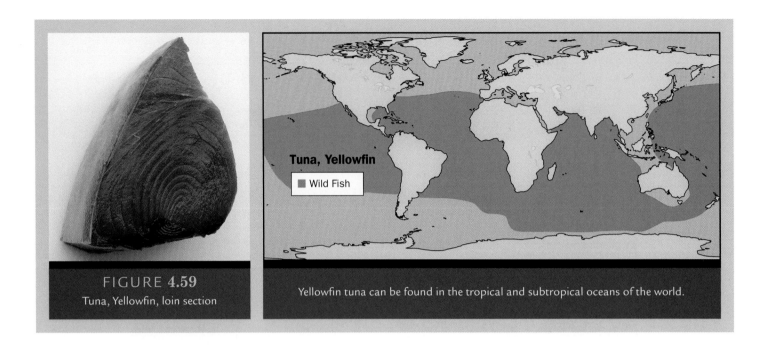

FIGURE 4.59
Tuna, Yellowfin, loin section

Yellowfin tuna can be found in the tropical and subtropical oceans of the world.

TUNA, YELLOWFIN (3 OZ SERVING)

Water: 60.34 g	Calories: 92	Total fat: 0.81 g	Saturated fat: 0.200 g
Omega-3: 0.237 g	Cholesterol: 38 mg	Protein: 19.87 g	Sodium: 31 mg

BONITO (SARD SARD)

Of all the commercial species of tuna, the bonito is the least valuable as a food fish. Found in both the Atlantic and Pacific, its meat is dark and very strong in flavor, which does not lend itself well to center of the plate applications. It can be identified by its distinctive body stripes that run on the bias along the body and a dark blue back and silvery belly. Average size ranges from 4 to 6 pounds. In Japanese cuisine, the meat is dried, smoked, and flaked, resembling wood shavings, which when combined with water and kelp, make the soup base/stock called *dashi*.

TUNA, BONITO (3 OZ SERVING)

Water: 60.34 g	Calories: 92	Total fat: 0.81 g	Saturated fat: 0.200 g
Omega-3: 0.237 g	Cholesterol: 38 mg	Protein: 19.87 g	Sodium: 31 mg

KATSU DASHI

Dashi is a Japanese stock used in a wide range of broths and soups in Japanese cuisine. Dashi consists of three main ingredients: water, kombu seaweed, and dried bonito flakes. Most of the bonito used for dashi are caught in the spring and summer months when they are lean. Once the fish begin to feed, their oil content is too high and the flavor becomes too strong for the delicate balance typical of dashi. Processing of the bonito begins by fabricating it into four separate loins, which are in turn blanched in salt water. Next it is hot-smoked and allowed to ferment at room temperature for up to two months. This fermentation develops the much sought-after umami flavor which, along with sweet, sour, salty, and bitter, are important elements in Japanese cuisine. After the bonito has fermented, it is very firm and quite dry. The last step in processing is to shave the loins into delicate flakes using a wooden rectangle-shaped device fitted with a razor-sharp blade. The best quality dashi is made from freshly shaved bonito, but many chefs purchase pre-packaged flakes of varying quality.

At first the process of making dashi appears simple, but like many things with a rich culinary history, it is more complex and detailed than most recipes suggest. The primary dashi is normally made just before service to ensure that the heat and water properly extract the correct flavor balance of the dried bonito flakes and the seaweed. Quality of the ingredients is very important because there are so few in dashi and any imperfections will be highlighted. When properly made and strained, the stock will be clear and flavorful and should be used for clear soup. A second, less powerful mixture can be made using the same kelp and bonito and this is best used for broths and soups when clarity is secondary. For a recipe for dashi, please see page 294.

TURBOT

- European (*Scophthalmus maximus*)
- Greenland (*Reinhardtius hippoglossoides*)

Turbot are one of the largest round boreal flat fish in the world. They are very dark grayish-green on one side and off-white on the other. Their shape is similar to a flounder except rounder and almost diamond-like. Several species are found throughout the Arctic, Atlantic, and Pacific oceans where they range in market size from 8 to 30 pounds. The European turbot is found throughout the Mediterranean, United Kingdom, and north throughout the Norwegian Sea.

Turbot landings are limited throughout the world due to previous overfishing and current catch restrictions, and demand keeps the market price high. The fish have excellent camouflage abilities but are easily caught as they lay stacked on top of each other on the ocean bottom. Turbot are farm-raised throughout Europe and Chile with average sizes ranging from 1 to 5 pounds.

QUALITY CHARACTERISTICS

From a culinary perspective, the European turbot is far superior in flavor and texture than its Greenland or Pacific cousin and demands a higher market price. European turbot has extremely white mild flesh, which is very firm. Only the Dover sole is more highly prized by chefs. True turbot should always be purchased whole, because fillets, especially those with the skin removed, can come from many impostors. Turbot have two distinct fillets on each side with the dark or backside fillets being thicker than the belly. Greenland turbot fillets are commonly available fresh or frozen and will be softer in texture.

Because this is such a prized flat fish, simplicity of preparation is very important. Highlight the fish's flavor and texture by limiting ingredients and choosing the correct cooking methods.

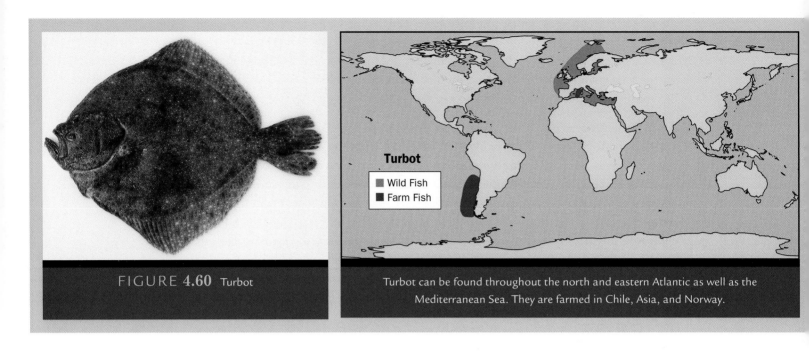

FIGURE 4.60 Turbot

Turbot can be found throughout the north and eastern Atlantic as well as the Mediterranean Sea. They are farmed in Chile, Asia, and Norway.

PREPARATION METHODS

- Bake
- Broil
- Poach
- Sauté
- Steam

SEASONAL AVAILABILITY

Year-round

TURBOT

Water: 65.41 g   Calories: 81   Total fat: 2.51 g   Saturated fat: 0.637 g
Omega-3: 0.51 g   Cholesterol: 41 mg   Protein: 13.64 g   Sodium: 128 mg

WOLFFISH (*ANARHICHAS LUPUS*)

Also called Atlantic catfish, this large fish has a blunt head, cylindrical-shaped body, and a very unattractive mouth with large sharp teeth used to feed on crustaceans and bottom fish. Dark bluish-gray to brown in color, it has dark vertical stripes on its sides and a long dorsal fin running down the length of its back. Its pectoral fins are large and fan-shaped and its anal fin runs from the center of its body to its tail.

Inhabiting the northern Atlantic, the wolffish is typically a bycatch of cod and is popular in Norway where it is farm-raised.

QUALITY CHARACTERISTICS

Underutilized in the United States possibly because of its appearance, wolffish has white meat that is slightly sweet, with a firm texture somewhere between cod and monkfish. It lends itself to a variety of cooking methods.

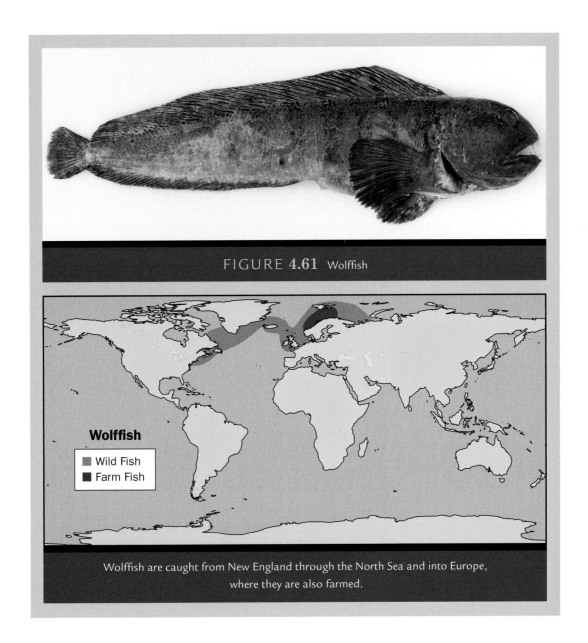

FIGURE 4.61 Wolffish

Wolffish are caught from New England through the North Sea and into Europe, where they are also farmed.

PREPARATION METHODS

- Bake
- Broil
- Fry
- Grill
- Poach
- Sauté
- Steam

SEASONAL AVAILABILITY

Year-round

WOLFFISH (3 OZ SERVING)

Water: 67.92 g Calories: 82 Total fat: 2.03 g Saturated fat: 0.310 g
Omega-3: 0.678 g Cholesterol: 30 mg Protein: 14.88 g Sodium: 72 mg

SHELLFISH IDENTIFICATION

CLASSIFICATIONS OF MARINE ANIMALS
ARTHROPODS (PHYLUM: ARTHROPODA)

Arthropods are one of the largest groups of invertebrate animals with over a million species. Of importance is the marine crustacean known for its jointed, partially flexible exoskeleton that protects the soft body. Because of the shell's rigidity, the animal grows by periodic molts, accomplished when a soft exoskeleton forms under the old one, which splits and cracks to allow exit. The new shell hardens rapidly and the animal has thus increased in size.

Bristles attached to their shells allow for taste, smell, sound, and touch, and their simple eyes give them sight. Arthropods have separate sexes, with the females laying fertilized eggs. In addition to the well-known culinary varieties, there are many smaller plankton species critical to the balance of the food chain.

COMMON ARTHROPODS
- Barnacles
- Crabs
- Lobster
- Shrimp

MOLLUSKS (PHYLUM: MOLLUSCA)

Mollusks are invertebrates consisting of seven classes including Aplacophora, Bivalvia, Cephalopoda, Gastropoda, Polyplacophora, Monoplacophora, and Scaphopoda. From a culinary and descriptive perspective, bivalves, gastropods, and cephalopods will be discussed.

BIVALVES

Comprising over 15,000 species, bivalves are mollusks with a two-part hinged wedge-shape shell that allows them to burrow into sandy bottoms. The shell encloses and protects the delicate body; the tough foot is used to dig and feed. Water is passed through a sheet-like meshed gill allowing it to breathe. To eat, food is trapped by excreted mucus that gets transferred into the mouth.

Some bivalves, such as scallops, can swim by quickly opening and closing their shells, whereas others use their strong foot to slowly propel along the bottom. Still others can attach themselves to hard objects and never move, and some, such as mussels, secrete fine threads that root themselves on stationary objects.

Reproduction for the majority of bivalves happens by spawning. Most have separate sexes producing only sperm or eggs, although some are hermaphroditic.

Common bivalves
- Clams
- Mussels
- Oysters
- Scallops

GASTROPODS

Most gastropods are characteristically known for their single shell, which can be spiral in shape like the snail or flat like abalone. Still others have no shell at all. The body consists of a foot that is used for propulsion, as well as a head, mouth, and eyes.

Common gastropods
- Abalone
- Conch
- Periwinkles
- Snails
- Whelk

CEPHALOPODS

Although soft and quite flexible, cephalopods are a specialized group of mollusks. These interesting creatures have highly developed vision and memory, and some have the ability to camouflage themselves.

A large bulbous mass surrounds the mantle, gills, body, and head, which contains the brain, organs, and mouth. Like most mollusks, cephalopods have a foot, which presents as a number of suction-cupped arms or tentacles surrounding its beak-shaped mouth.

Squids have soft, flexible internal shells called quills, whereas octopuses have no shell at all. Cephalopods have an ink gland, or sac, that can be released when threatened. This ink is edible and has many culinary applications.

Common cephalopods
- Cuttlefish
- Octopus
- Squid

ECHINODERMS (PHYLUM: ECHINODERMATA)

Found only in salt water, this group of radial-shaped animals is the least important from a culinary standpoint because they lack significant amounts of edible meat. A layer of skin covers the soft bodies of echinoderms. Some have spines that are fixed or flexible and all species vary in size.

COMMON ECHINODERMS
- Sea cucumber
- Sea urchin

ABALONE (FAMILY: HALIOTIDAE)

Abalones are large univalve mollusks or gastropoda, which have a one-piece flat shell that is rounded or oval-shaped like an ear. Averaging 4 pounds, they can grow up to 8 pounds. The shell contains rows of respiratory pores; the foot is extremely strong allowing it to clamp onto rocks. Worldwide there are approximately 100 species, and in North America there are nine.

- Black
- Flat
- Green
- Pink
- Pinto
- Red
- Threaded
- Western Atlantic
- White

Differences include variations in shell size, color, number of pores, and habitat. Most species are found in subtidal zones at depths from 20 to 120 feet with the majority located in the Pacific Ocean.

Because of the delicacy of the meat and the beauty of the shell, abalones have been overharvested for many years. Man and the sea otter have depleted the stock in the Pacific to record lows, precipitating regulations in California banning the sale of fresh and frozen precuts out of state.

QUALITY CHARACTERISTICS

Abalone is extremely tough and must be tenderized. The steak is white with a beautiful flavor, and should be sliced thinly against the grain. The entire flesh of the abalone can be eaten but typically in the United States, only the main muscle is consumed. In Japan the gonads are a delicacy but preparations for this must be done immediately upon shucking. Prolonged cooking will toughen the meat so limited cooking is preferred, or it can be eaten raw. Abalone is farm-raised and market forms include fresh, frozen, smoked, and canned. The shell is valued for mother of pearl, used in furniture inlays and jewelry production and has historically been as valuable as the meat.

FIGURE 5.1 Abalone

Wild caught abalone can be found in Mexico, Central America, Asia, and Australia. Abalone is also farmed throughout the world.

PREPARATION METHODS

- Pan fry
- Raw
- Sauté
- Stir fry

SEASONAL AVAILABILITY

Year-round

ABALONE (3 OZ SERVING)

Water: 63.38 g	Calories: 89	Total fat: 0.65 g	Saturated fat: 0.127 g
Omega-3: 0.042 g	Cholesterol: 72 mg	Protein: 14.54 g	Sodium: 265 mg

CLAMS

Clams are bivalve mollusks with two separate shells locked together by a pair of strong adductor muscles and a foot that helps it to bury itself up to 2 feet in the sandy bottom.

Clams feed and breathe by drawing in and expelling water through tubes or siphons. They range in size from miniscule to over 3 feet across. There are over 12,000 species found worldwide.

Clams can be affected by toxins from marine algae called red tide, and from sewage and human waste. Commercial clam fishers and government agencies monitor the beds and have an excellent safety program in place to ensure a wholesome product. The United States Department of Agriculture (USDA) recommends not eating raw seafood especially if you are in a high-risk group, and also recommends that those choosing to consume it raw should freeze it first, although freezing alone will not eliminate all contaminants.

CLAM, GEODUCK (*PANOPEA GENEROSA*)

Pronounced "gooey duck," this large clam native to the Pacific coast of North America is one of the oldest living organisms and has a life expectancy of well over 100 years. Also referred to as elephant trunk clam or king clam, its name is thought to be of Native American origin meaning to "dig deep." Geoduck clams are found buried in up to 3 feet of sand and silt where they feed on phytoplankton by sticking their long necks up and out into the water. Because their body and siphons are so large, the shells are always open. A rarity in Asia, geoduck clams are increasingly being farm-raised

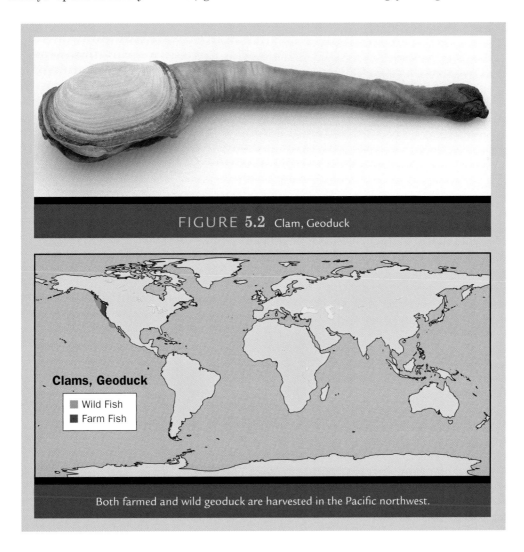

FIGURE 5.2 Clam, Geoduck

Both farmed and wild geoduck are harvested in the Pacific northwest.

throughout the Pacific Northwest, especially around the Puget Sound. Geoduck clams are also harvested by underwater divers using high-pressure hoses to aid in digging.

QUALITY CHARACTERISTICS

A delicacy in Asia, they are eaten as an element in Chinese hot pot, or raw as sashimi. In the United States, they are ground up and used in chowders, made into fritters, or pounded out and sautéed or pan fried.

PREPARATION METHODS

- Fry
- Pan fry
- Raw
- Sauté
- Stew

SEASONAL AVAILABILITY

Year-round

CLAM, GEODUCK (USED MIXED CLAM SPECIES DATA)

Water: 69.55 g   Calories: 63   Total fat: 0.82 g   Saturated fat: 0.08 g
Omega-3: 0.241 g   Cholesterol: 29 mg   Protein: 10.85 g   Sodium: 48 mg

CLAM, HARD-SHELL (*MERCENARIA MERCENARIA*)

Also marketed as *quahog* or by names relating to specific sizes, hard-shell clams are found in the waters from Canada south through Florida with concentrations around the Chesapeake Bay. They are harvested with hydraulic dredges and rakes and are also farm raised. They have a thick gray to tan shell with raised concentric growth lines.

MARKET FORMS FOR HARD-SHELL CLAMS

- Canned
- Freshly shucked in tubs
- Frozen
- Individually quick-frozen (IQF) half shells
- Live in mesh bags
- Stuffed half shells

SIZES AND MARKET NAMES OF HARD-SHELL CLAMS FROM SMALLEST TO LARGEST

- Littlenecks
- Topnecks
- Cherrystones
- Chowder clams

QUALITY CHARACTERISTICS

Hard-shell clams are the best of all species to be served raw on the half shell. They have a sweet flavor and delicate texture. When they are cooked, they are white to flesh colored and remain sweet and very juicy with a mild briny flavor reminiscent of ocean mist. Live clams should be purchased fresh and be tightly closed and kept in a

FIGURE 5.3 Clam, Hard-shell

Hard-shell clams are both dredged and farmed along the eastern seaboard from Florida to Canada.

cold, damp environment when out of water. Any opened or cracked clams should be discarded. They suffocate if tied up in plastic bags or sealed in containers.

PREPARATION METHODS

- Bake
- Poach
- Steam
- Raw
- Fry
- Sauté
- Stir fry

SEASONAL AVAILABILITY

Year-round

CLAM, HARD-SHELL (3 OZ SERVING)

Water: 69.55 g	Calories: 63	Total fat: 0.82 g	Saturated fat: 0.08 g
Omega-3: 0.241 g	Cholesterol: 29 mg	Protein: 10.85 g	Sodium: 48 mg

CLAM, RAZOR

- Atlantic species (*Siliqua costata*)
- Pacific species (*Siliqua patula*)

The razor clam is a bivalve found along the Atlantic Coast from Canada to the Carolinas and in the Pacific from Alaska to California. The Atlantic species is the smaller of the two, measuring about 2-1/2 inches long; its Pacific cousin reaches a length of 7 inches. Both clams are oblong in shape and resemble an old-style straight edge razor or jack-knife, its other common name. The Atlantic clam is white with colorations of purple and brown whereas the Pacific clam is white with colorations of yellow and brown.

QUALITY CHARACTERISTICS

The meat of these clams is off-white and creamy but tough in texture and can be eaten on the half shell only when very small. Most of the Pacific razor clams are made into fritters whereas the Atlantic varieties are typically steamed.

SEASONAL AVAILABILITY

Year-round

CLAM, RAZOR (3 OZ SERVING)

Water: 69.55 g	Calories: 63	Total fat: 0.82 g	Saturated fat: 0.08 g
Omega-3: 0.241 g	Cholesterol: 29 mg	Protein: 10.85 g	Sodium: 48 mg

FIGURE 5.4 Clam, Razor

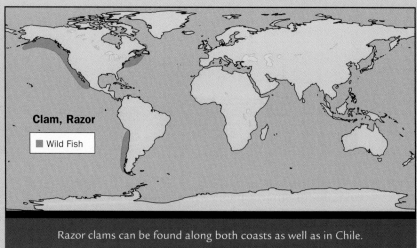

Razor clams can be found along both coasts as well as in Chile.

CLAM, SOFT-SHELL (*MYA ARENARIA*)

The soft-shell clam is a native of the East Coast and British Isles. Named for its soft and brittle shell, a majority of the harvesting is done in Chesapeake, Massachusetts, and Maine. Other common names are belly clam, Ipswich, longneck, and steamer. The soft-shell clam has an elongated form and is brushed white with a market size of 2 to 3 inches in length. The clam's neck or siphon protrudes out and the shell does not close completely, making them more delicate and perishable than hard-shell clams. Careful harvesting is done with rakes in sandy soil, so purging helps to eliminate grit.

PURGING SOLUTION

Water	1 gallon
Kosher salt	3 tablespoons
Corn meal	3 tablespoons

Method

Mix together the water, salt, and cornmeal. Cover clams and change the water every 45 minutes to keep the water oxygenated. Repeat, as needed, depending on the harvesting environment.

QUALITY CHARACTERISTICS

Live clams should be active and their necks should move when handled. Always purchase them live and evaluate each one to ensure freshness and use as soon as possible. The flavor tends to be mild and briny, which contrasts nicely with sauces. Choose other clams to eat on the half shell because these are best simply prepared by steaming. Wash the clams and steam in an inch or two of wine, water, or stock flavored with herbs, garlic butter, salt, and pepper in a covered pot for 4 to 6 minutes. It is a good idea to remove the dark outer portion of the neck before serving. They can also be removed from the shells and fried.

CLAM, SOFT-SHELL (3 OZ SERVING)

Water: 69.55 g	Calories: 63	Total fat: 0.82 g	Saturated fat: 0.08 g
Omega-3: 0.241 g	Cholesterol: 29 mg	Protein: 10.85 g	Sodium: 48 mg

FIGURE 5.5 Clams, Soft-shell

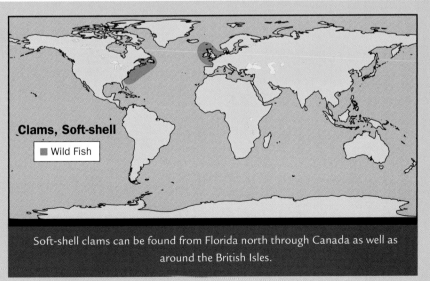

Soft-shell clams can be found from Florida north through Canada as well as around the British Isles.

FIGURE 5.6 Clams, Surf — Surf clams are found around the eastern Atlantic from the Carolinas to Canada.

CLAM, SURF

Averaging 4 to 8 inches in length, they are triangular and slightly oval in shape and are light brown to gray in color. Harvested year-round by large vessels from Canada to the Carolinas, they are one of the most commercially important of all clam species.

QUALITY CHARACTERISTICS

Because of their large size, surf clams are coarse and chewy. The tongue is typically removed and breaded into clam strips, while the rest of the body or juice is ground or sold for soup and sauces. Because of its product versatility, surf clams are made into a wide range of products.

SURF CLAM PRODUCTS

- Breaded clam strips
- Chowder
- Clam cakes
- Clam juice
- Clam stuffing
- Fresh chopped meat
- Minced clams

SEASONAL AVAILABILITY

Year-round

CLAM, SURF (3 OZ SERVING)

Water: 69.55 g Calories: 63 Total fat: 0.82 g Saturated fat: 0.08 g
Omega-3: 0.241 g Cholesterol: 29 mg Protein: 10.85 g Sodium: 48 mg

CONCH, QUEEN (*STROMBUS GIGAS*)

Nothing epitomizes the tropics more than the beautiful Queen Conch, a gastropod with a distinctive conical pink shell that is sold to tourists from the Florida Keys and throughout the Caribbean. Similar to its smaller northern cousin the whelk, conch is gathered by hand throughout the region. The large shell surrounds the edible muscle, which can be loosened by tapping away the third spiral down from the top of the shell. This will allow access for a special corkscrew-like tool or knife to separate the meat from the shell, for easy removal.

QUALITY CHARACTERISTICS

Like other gastropods, conch is tough and must be tenderized. The first step in fabrication is to remove the outer skin by carefully slicing it off with a sharp knife and is easiest when the conch is fresh. Another method is to butterfly the entire piece open and pound it with a mallet until thin and tender. Other methods of preparation involve blanching the meat in boiling water several times before braising, and it is often tenderized using a pressure cooker.

Along the beaches of the Bahamas conchs are sliced thin and eaten raw with a dousing of lime. They are also ground up and made into chowder or mixed in a batter and made into fritters. Cracked conch is prepared by butterflying the meat into steaks, then breading and deep frying it. Typically, it is served with a spicy tomato Creole sauce.

Similar in flavor and texture to a clam, the conch has a sweet distinct taste that is often paired with hot sauce and island spices. Conch is being overfished, and to preserve the species, many areas have imposed strict harvesting quotas. It has been banned commercially in Florida, but imports of farm-raised conch are available from Turks and Caicos in sizes ranging from a tender several inches to large full-grown adults. Market forms in the United States include frozen 5-pound boxes or fresh local or farm-raised varieties.

PREPARATION METHODS

- Bake
- Braise
- Deep fry
- Grill
- Pan fry
- Sauté
- Stew

FIGURE 5.7 Conch, Queen

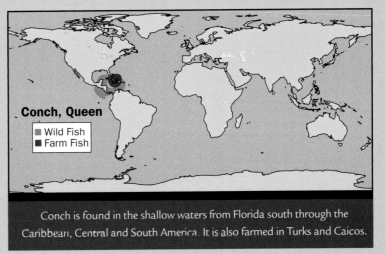

Conch is found in the shallow waters from Florida south through the Caribbean, Central and South America. It is also farmed in Turks and Caicos.

SEASONAL AVAILABILITY

Farm-raised and frozen conch are available year-round.

CONCH (1 OZ SERVING)

Water: 11.80 g Calories: 22 Total fat: 0.20 g Saturated fat: 0.063 g
Omega-3: 0.102 g Cholesterol: 11 mg Protein: 4.47 g Sodium: 26 mg

CRAB

The North American continent is graced with an abundance of delicious crabs available year-round in many forms including cooked, pasteurized, and frozen. The versatile meat is sometimes labor intensive to remove but worth the work especially when paired with simple accompaniments such as citrus, butter, and herbs.

PREPARATION METHODS

- Boil
- Fry
- Sauté
- Stir fry
- Broil
- Pan Fry
- Steam

CRAB, BLUE (*CALLINECTES SAPIDUS*)

The blue crab is found throughout the Western Atlantic and Southern Asia and has successfully been established in other areas throughout the world. It can live in varying degrees of salinity and in the United States is concentrated around the Chesapeake Bay and Gulf of Mexico. It has an oval-shaped shell that is white to tan on the underside with shades of brown, blue, and green on top. The back pair of legs are flat, contributing to their Latin name Callinectes, meaning "beautiful swimmer." Available from spring through fall, their average market size measures about 4 to 5 inches across the shell. Males can be identified by their blue claws; females' claws have orange tips.

SEASONAL AVAILABILITY

April to November

CRAB, BLUE (3 OZ SERVING)

Water: 67.17 g Calories: 74 Total fat: 0.92 g Saturated fat: 0.189 g
Omega-3: 0.403 g Cholesterol: 66 mg Protein: 15.35 g Sodium: 249 mg

CRAB, SOFT-SHELL

All crabs lose their shells in order to grow, and at this point the blue crab is called a soft-shell. Its small size and delicate meat make it one of the most anticipated culinary events of the year. Through May and July, commercial crabbers keep hard-shell blue crabs in floating containers and monitor their development. When they finally pop their shell, their body size expands and they are immediately removed from the water and sold as soft-shell crabs.

FIGURE 5.8 Crabs, Blue

Blue crabs are found in the Atlantic from the northern areas of South America to New York as well as throughout Asia.

QUALITY CHARACTERISTICS

Although it is labor intensive to pick, fresh blue crabmeat is sweet, rich, and delicate in texture and flavor and is worth the effort. Purchase only fresh crabs that are very active; any dead crabs should be discarded. Boil or steam them immediately. Claw meat is off-white in color and has a slightly different texture and flavor than the body meat. Blue crabmeat is available in pasteurized 1 pound tubs, and soft-shells are available fresh in season or dressed and frozen year-round. Soft-shell crabs should have their bottom aprons taken off and their gills removed before cooking (refer to soft-shell crab fabrication).

SEASONAL AVAILABILITY

Available from April to September with concentrations in June and July.

CRAB, BLUE (SOFT-SHELL) (3 OZ SERVING)

Calories: 87 Total fat: 1.1 g Saturated fat: 0.2 g Omega-3: 0.3 g
Cholesterol: 78 mg Protein: 18.1 mg Sodium: 293 mg

CRAB, KING (*PARALITHODES CAMTSCHATICUS*)

Also referred to as the Alaskan king crab, this giant can weigh in at over 20 pounds but market size tends to be in the 8 to 10 pound range with about a 25 percent meat yield. Caught in large pots or traps in the icy water off the coast of Alaska, the Bering Sea, and around Japan, they are extremely perishable and the majority of the catch is cooked soon after landing and frozen or canned. Shell coloration is determined by species of which there are three: red, blue, and brown with the red king crab being the most sought after and identified by its brilliant orange red to rust shell color. The legs are very long, have pointed appendages, and contain the majority of the meat. Body meat is mostly picked, canned, and cooked immediately after landing. Strict fishing quotas have rebuilt the once decimated stock.

FIGURE 5.9 Crab, King crab legs

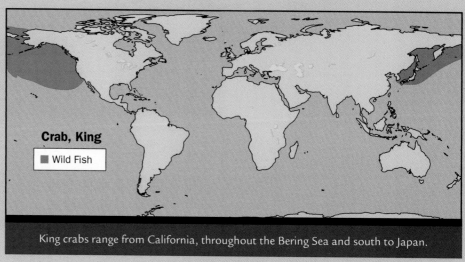

King crabs range from California, throughout the Bering Sea and south to Japan.

QUALITY CHARACTERISTICS

Sold in many forms, the most popular being cooked legs and leg sections. The premium and most tender of all the meat comes from the upper leg sections. All the meat is buttery rich and firm with a sweet nutty flavor. Body sections are more delicate and flaky than claw meat. Because the claws are sold frozen and are difficult to package, it is very important to avoid purchasing claws that have been freezer burned due to thawing or prolonged storage. Thaw crab legs slowly in the refrigerator and use immediately because they can become waterlogged and absorb refrigerator flavors easily. Most of the crabmeat is precooked, therefore it is important to only reheat to proper internal temperatures before consuming. Do not cook for a prolonged period of time or the delicate flavor and texture will be ruined.

SEASONAL AVAILABILITY

September to December

CRAB, KING (3 OZ SERVING)

Water: 67.63 g   Calories: 71   Total fat: 0.51 g
Cholesterol: 51 mg Protein: 15.55 g   Sodium: 59 mg

CRAB, DUNGENESS (*CANCER MAGISTER*)

This dark brown hard-shell crab gets its name form the town of Dungeness, Washington, located northwest of Seattle and near Port Angels where the annual Dungeness festival is held each autumn. Harvested from Southern California to Alaska, the Dungeness has a wide hard shell with ten legs, including a pair of claws used for feeding on crustaceans and small fish. The Dungeness is an important commercial crab throughout California and the Pacific Northwest and can be found at both the famed Pike Place market in Seattle and Fisherman's Wharf in San Francisco.

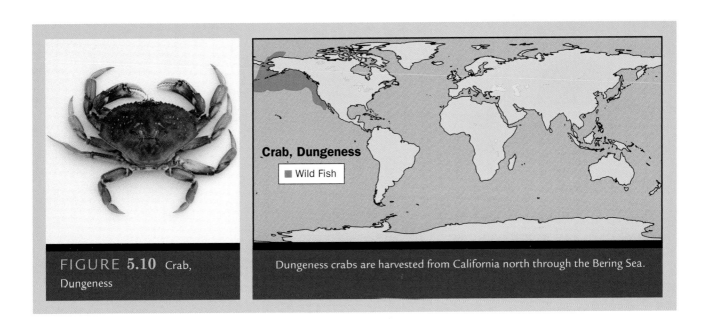

FIGURE 5.10 Crab, Dungeness

Dungeness crabs are harvested from California north through the Bering Sea.

QUALITY CHARACTERISTICS

Although only about a quarter of the crab's body weight is edible meat, it is prized for its delicate flavor and texture. Available in markets live or cooked, which turns it a rusty orange color, it must measure 6-1/4 inches across to be harvested. Cooked meat can be utilized in soups, chowders, salads, and crab cakes, along with many other preparations. A nutcracker or hammer is used to break the shell, after which the meat can be picked out with a shrimp fork.

SEASONAL AVAILABILITY

Year-round

CRAB, DUNGENESS (3 OZ SERVING)

Water: 67.30 g	Calories: 73	Total fat: 0.82 g	Saturated fat: 0.112 g
Omega-3: 0.335 g	Cholesterol: 50 mg	Protein: 14.80 g	Sodium: 251 mg

CRAB, JONAH (*CANCER BOREALIS*)

Found along the East Coast from Canada to Florida, Jonah crabs are caught in pots and as bycatch in lobster traps. They have an oval shape and are rust to brown on top and have an off-white to yellow belly; measuring about 6 inches across. Jonah crabs are available year-round.

QUALITY CHARACTERISTICS

The flavor of the Jonah crab is mild compared to the blue or king crab but the meat is firm, buttery, and sweet; the cooked meat is white and brown. The claws are especially prized. It is often less expensive than blue crab, hence suitable and economical in any recipe calling for crabmeat. Market forms include pasteurized meat, fresh live crabs, and a variety of processed claw products. Live crabs should be active and should be cooked immediately.

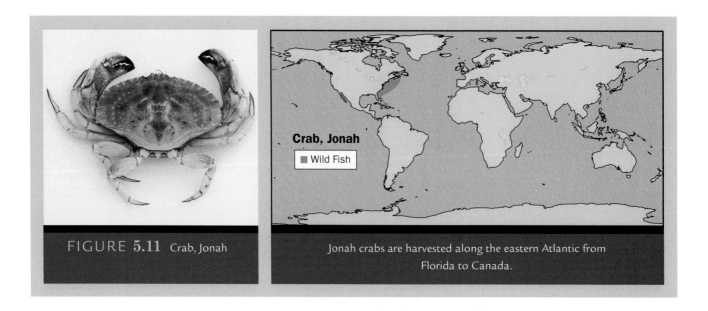

FIGURE 5.11 Crab, Jonah

Jonah crabs are harvested along the eastern Atlantic from Florida to Canada.

SEASONAL AVAILABILITY

Year-round

CRAB, JONAH (3 OZ SERVING)

Calories: 81   Total fat: 1.62 g   Saturated fat: 0.14 g
Cholesterol: 66 mg   Protein: 13.77 g   Sodium: 234.6 mg

CRAB, SNOW (*CHIONOECETES OPILIO*)

Several similar species are marketed as snow crabs and common names are queen crab, spider crab, and tanner crab. Snow crabs have eight legs, two claws, and are orange to red in color. They range in size from 1 to 6 pounds and are mostly found in the Northern Pacific, although certain species are trapped in the Northern Atlantic.

QUALITY CHARACTERISTICS

Snow crabmeat is mild in flavor and has a stringier texture than the traditional lump or firm consistency of other species. Because they are so perishable all snow crabs are cooked and available frozen.

SEASONAL AVAILABILITY

Frozen meat and legs are available year-round.

CRAB, SNOW (3 OZ SERVING)

Calories: 77   Total fat: 1.11 g   Saturated fat: 0.14 g
Omega-3: 0.374 gm   Protein: 15.64 g

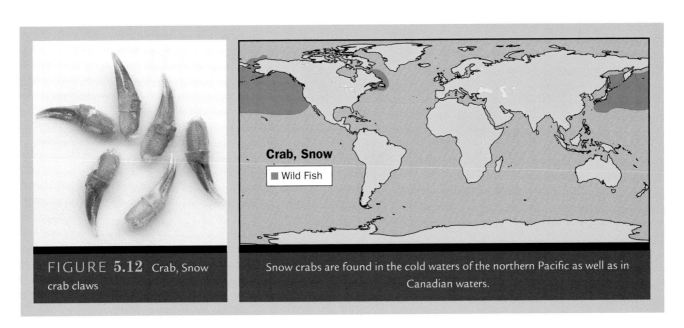

FIGURE 5.12 Crab, Snow crab claws

Snow crabs are found in the cold waters of the northern Pacific as well as in Canadian waters.

CRAB, STONE (*MENIPPE MERCENARIA*)

Stone crabs get their name from their hard, oval stone-like shell. They are fished from the Carolinas through the Gulf of Mexico. Easily identified by their black-tipped claws,

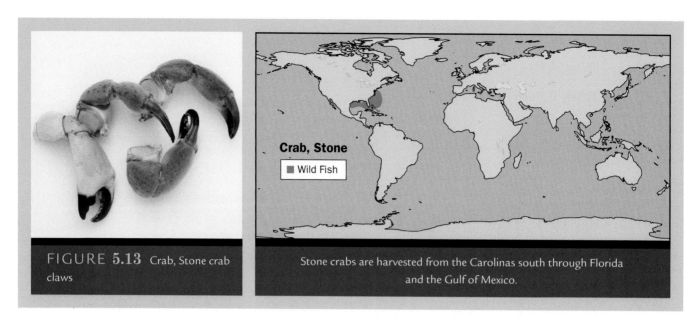

FIGURE 5.13 Crab, Stone crab claws

Stone crabs are harvested from the Carolinas south through Florida and the Gulf of Mexico.

they average about 5 inches across; the body color is purplish brown to red. Stone crabs are caught in traps; when brought on board the front portion of the claws are removed and the crabs thrown back. Claw regeneration takes about 1-1/2 years; the crabs have a life span of about 8 years. Fishing live crabs is illegal in Florida.

QUALITY CHARACTERISTICS

Market forms include cooked and frozen claws. The cooked meat is sweet and firm, closer in texture to a lobster than to a blue or king crab. Shells are hard and will need to be cracked opened with a nutcracker. As with all frozen crab, proper thawing in the refrigerator will ensure a quality product.

SEASONAL AVAILABILITY

October to May

CRAB, STONE (3.5 OZ SERVING)

Calories: 71   Total fat: 0 g   Saturated fat: 0 g
Cholesterol: 53 mg   Protein: 17.6 g   Sodium: 353 mg

CUTTLEFISH (SEPIA OFFICINALIS)

Cuttlefish look similar to squid but can be distinguished by their flat, wide and substantial body, small fins, and compact shape. Ranging in size from just a few inches to several feet, they are light brown with a striped membrane over their mantel which should be removed when cleaning. They are common throughout the Eastern Atlantic, Mediterranean, India, and the Pacific.

QUALITY CHARACTERISTICS

Because of their wider and more compact shape, cuttlefish are thicker than squid. Their ink sac is also larger, but is often crushed in the capture. Cooked meat is white, sweet, and very tender when cooked quickly using high heat such as grilling or frying. They can also be braised or stewed.

PREPARATION METHODS

- Boil
- Broil
- Fry
- Pan Fry
- Sauté (picked meat or soft-shells)
- Steam
- Stir fry

SEASONAL AVAILABILITY

Year-round

CUTTLEFISH (3 OZ SERVING)

Water: 68.48 g   Calories: 67   Total fat: 0.59 g   Saturated fat: 0.100 g
Omega-3: 0.179 g Cholesterol: 95 mg   Protein: 13.80 g   Sodium: 316 mg

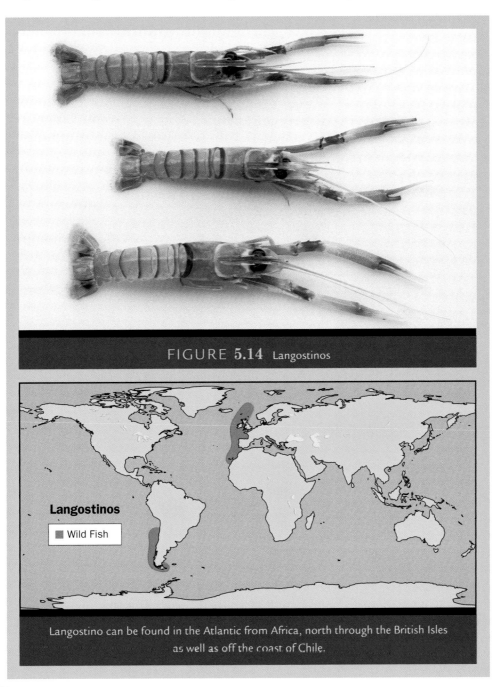

FIGURE 5.14 Langostinos

Langostinos
■ Wild Fish

Langostino can be found in the Atlantic from Africa, north through the British Isles as well as off the coast of Chile.

LANGOSTINO (*CERVIMUNIDA JOHNI, MUNIDA GREGARIA, PLEURONCODES MONODON, NEPHROPS NORVEGIUS*)

Langostino is the name given to a variety of crustaceans with a similar appearance to a small lobster. They are confusing because of the variety of species and the many names and marketing terms associated with them including prawns, Dublin bay prawns, scampi, and lobsterette among others. Inhabiting the deep waters of continental shelves and found from Iceland to Morocco and throughout the Adriatic, their shell colorations vary among species but typically they are beautiful yellowed gold to red with similar colored bands on their claws. They look like a cross between shrimp and lobster with long pinchers, a fanned tail, and long spiny legs. Typical body length is between 8 to 10 inches. Imported varieties are available fresh or frozen from Europe where they have been enjoyed for centuries.

QUALITY CHARACTERISTICS

The meat is firm and more like shrimp or lobster than crab. Its flavor is sweet and buttery, with aromas of shrimp when cooked. It is available fresh, and in a variety of IQF products. Quick cooking methods are the best bet for the delicate tails or they toughen and get dried out.

PREPARATION METHODS

- Broil
- Poach
- Steam
- Grill
- Sauté
- Stir fry

SEASONAL AVAILABILITY

Year-round

LANGOSTINO (3 OZ SERVING)

Calories: 63   Total fat: 1.36 g   Saturated fat: 0.27 g
Omega-3: 0.493 g   Protein: 6.97 g

LOBSTER

Most people enjoy a hot buttery lobster. Children are fascinated by these strange sea creatures that often are piled on top of each other in the bubbly cold glass tank at the supermarket or restaurant. Adults savor its succulent meat. Lobsters are unique, one of the few foods that must remain alive until cooked and whose consumption requires the diner to utilize tools and a bib.

Lobsters live on the bottom of the ocean floor, securing themselves amongst rocks, and feeding on many varieties of marine life. They are invertebrates that mature by molting their hard exoskeleton, live very long lives, and grow to over 40 pounds. Baited traps are invariably used to ensnare them, except in the tropics and subtropics where divers search the ocean floor by hand and snare for the spiny varieties.

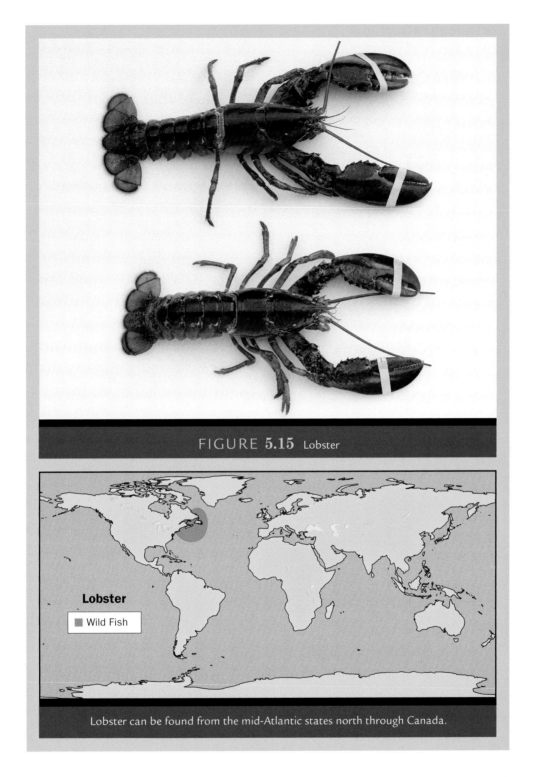

FIGURE 5.15 Lobster

Lobster can be found from the mid-Atlantic states north through Canada.

There are dozens of lobster species worldwide with the most common being the Maine, or American, lobster, and the spiny lobster, found in tropical and subtropical regions.

LOBSTER, AMERICAN (*HOMARUS AMERICANUS*)

Once so plentiful that they were used as fertilizers and fish bait, it wasn't until modern refrigeration and transportation that this crustacean crawled out of obscurity. Now they are prized and offered as high-end indulgence items, flown overnight from Boston and other cities to the restaurants of the world.

Contrary to what most people believe, lobsters are nearly 100 percent edible. The tails, claws, and legs yield a deliciously moist, firm meat. Many people enjoy the internal organs including the roe and tomalley (liver and pancreas), which is edible but contains possible environmental toxins, especially in older species. Shells can be roasted until brittle, before being cooked down, pureed, and strained to make delicious stocks, sauces, or bisque.

Lobsters are harvested from cold New England waters from Labrador to North Carolina with concentrations in Maine, Massachusetts, and Canada. Caught in traps baited with herring, lobsters are caught live and are put in holding tanks and shipped worldwide. Their claws are secured with rubber bands immediately after removal from the traps. Legal size limits are 3-1/2 to 5 inches from the back of their eyes to the end of the caripace. Once considered the bug of the sea, they are an important industry throughout New England and Canada and trapping them is strictly controlled.

Live lobster can be held under refrigeration for several days as long as they are in a moist, not wet environment. Covering them with wet newspaper or seaweed or in a salt water holding tank works best. Do not place them in fresh water or on ice.

Periodically, lobsters lose their shells and molt to grow. During this regrowth stage, the new shell will be soft, the meat wet and spongy, and they should be avoided. Yields vary depending on molting, feeding, and holding cycles but should be in the 20 to 30 percent range for 1-1/2 pound lobsters. Coloration depends on water conditions, but most are mottled rust, brown, green, and blue, turning a fiery red when cooked.

Edible meat can be found in the tail, claws, knuckles, and feet but requires some work to extract. Lobsters are best steamed or boiled, although they can be split in half and baked, broiled, or grilled. Scientists argue about the ability of the lobster to feel pain, but a knife driven between the eyes will be sufficient to end any suffering.

Determining doneness is confusing for some, with many people mistakenly thinking they should be cooked until a white protein-like substance comes out of the shell. At this point the lobster is overcooked. The best way to determine whether a lobster is done is to test it with an instant read thermometer inserted in the tail; when it reaches between 140° to 145°F/60° to 63°C, it's cooked. Because the meat is encapsulated in a hard shell, it will continue to cook once removed from the heat, so plan accordingly. Cooked lobsters can be shocked in ice water to stop the cooking, making it easier to remove the meat from the shell. Raw meat is virtually impossible to remove because it is jelly-like and very wet. To remove undercooked meat to be used in other preparations, blanch lobsters for a minute or two to firm the meat for easier extraction. In this method it is imperative that it be used immediately. Lobsters should always be handled live or cooked. Dead lobsters quickly deteriorate and should be avoided.

QUALITY CHARACTERISTICS

Cooked lobster meat is white and red in color, with a mild sweet flavor and firm texture. The tail meat is the most fibrous followed by the claws, knuckles, and feet, which are sweet and delicate.

Available year-round out of New England and Canada, lobster prices vary based on time of year, supply and demand, with the lowest prices typically in the late summer. Because lobsters are so commercially valuable, farming has been studied as a way to lessen exploitation. However, due to their aggressive and cannibalistic nature and slow growth rate, this procedure is not commercially viable. Depending on water conditions and environment, it can take a lobster 5 to 10 years to reach 1-1/2 pounds. Although lobsters grow faster in warmer water, they consume much more food and are more susceptible to disease, increasing production costs for those that are farm-raised.

PREPARATION METHODS

- Bake
- Boil
- Broil
- Grill

SEASONAL AVAILABILITY

Available year-round with seasonal availability from July to October.

LOBSTER, MAINE (3 OZ SERVING)

Water: 65.25 g Calories: 76 Total fat: 0.77 g Saturated fat: 0.153 g
Omega-3: 0.071 g Cholesterol: 81 mg Protein: 15.98 g Sodium: 252 mg

LOBSTER, SPINY (*PANULIRUS ARGUS*)

There are many species of spiny lobsters found throughout the world. Warm water tails come from the Caribbean and South America and are reddish brown, whereas cold water varieties are from Australia and New Zealand and are more orange in color. The spiny lobster is clawless with long antennae and the tail contains most of the edible meat. Average tail size weighs from 8 to 16 ounces. Most lobsters are caught in traps or

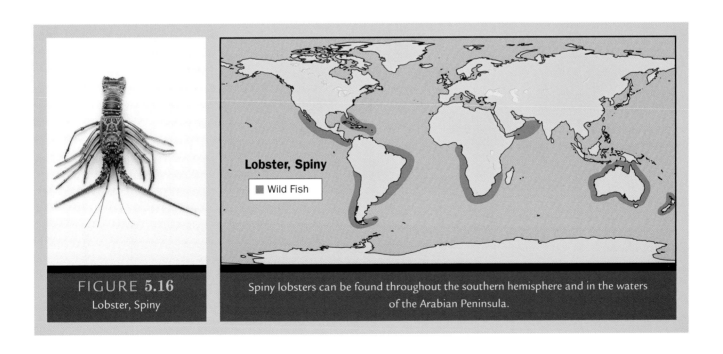

FIGURE 5.16
Lobster, Spiny

Spiny lobsters can be found throughout the southern hemisphere and in the waters of the Arabian Peninsula.

by divers. Due to overfishing, most countries have limited seasons; world supply varies throughout the year.

QUALITY CHARACTERISTICS

Very firm in texture and mild in lobster flavor, the cold water spiny lobster is generally considered superior to the warm water variety. The majority of the tails are available frozen. The shell is harder than the Maine lobster and the flavor is not as sweet. Most warm water tails are ice glazed which can add excess weight to an already expensive product. Avoid meat that is off-color and ensure that the tails have been properly cleaned of the digestive vein prior to freezing. Tails should be slowly defrosted in the refrigerator and not under running water, which will leach the flavor. Cook spiny lobster quickly or they will become tough and grainy in texture. Inserting a thermometer into the tail meat is a good way to judge doneness. Correct internal temperature should range between 140° to 145°F/60° to 63°C.

SEASONAL AVAILABILITY

Frozen tails are available year-round.

LOBSTERS, SPINY (3 OZ SERVING)

Water: 62.96 g	Calories: 95	Total fat: 1.28 g	Saturated fat: 0.201 g
Omega-3: 0.408 g	Cholesterol: 60 mg	Protein: 17.51 g	Sodium: 150 mg

MUSSEL, BLUE (*MYTILUS EDULIS*)

The blue mussel can be found worldwide where they are harvested from intertidal zones. Increasingly more common are farm-raised mussels, which grow on ropes or mesh structures suspended from a series of floating rafts. The farm-raised varieties can be distinguished by their thin dark shells and clean appearance. Wild mussels will typically have a beard or byssus located near the hinge, which is used to anchor itself onto rocks and pilings. Their shell will be very thick and shiny.

QUALITY CHARACTERISTICS

Cultivated mussels are generally harvested between 2 and 3 inches and have a sweet meaty flavor, medium texture, and are very nutritious. Their shells should be closed or show movement when tapped. Discard any dead or broken mussels immediately. Meat color varies from light brown to orange depending on habitat. Mussels should be received and stored in open meshed sacks and used immediately.

PREPARATION METHODS

- Bake
- Broil
- Sauté
- Smoke
- Steam

MUSSEL, BLUE (3.5 OZ SERVING)

Water: 80.58 g	Calories: 86	Total fat: 2.24 g
Saturated fat: .425 g	Protein: 11.9 g	Sodium: 286 mg

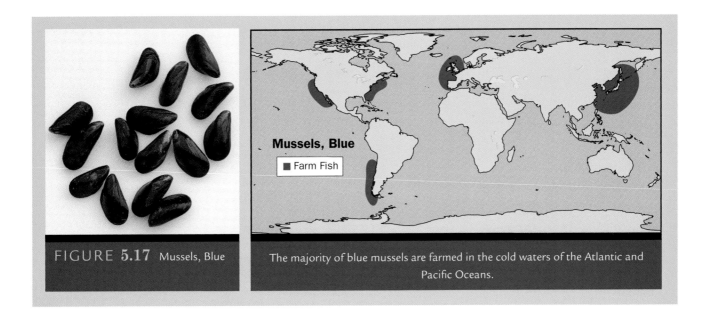

FIGURE 5.17 Mussels, Blue

The majority of blue mussels are farmed in the cold waters of the Atlantic and Pacific Oceans.

MUSSELS, GREEN LIP (*PERNA CANALICULUS*)

Originating in New Zealand, this farm-raised variety is strictly controlled and enjoys the pristine water conditions of the region. Average market size is between 3 and 4 inches and they can be identified by their long, beautiful shiny green and brown shell.

QUALITY CHARACTERISTICS

Available in myriad forms, green lip mussels are sweet and plump and considerably larger than the domestic blue mussel. Because they are cultivated, they are consistent in flavor, very clean, and devoid of sand or grit. When purchased fresh, the shell will often be open. As with all live shellfish, the mussel should react to touch or it is dead and should be discarded.

PREPARATION METHODS

- Bake
- Broil
- Sauté
- Smoke
- Steam

SEASONAL AVAILABILITY

Year-round

MUSSEL, GREEN (3.5 OZ SERVING)

Water: 68.49 g	Calories: 73	Total fat: 1.90 g	Saturated fat: 0.361 g
Omega-3: 0.655 g	Cholesterol: 24 mg	Protein: 10.12 g	Sodium: 243 mg

SCALLOPS

Scallops are a bivalve mollusk found in oceans worldwide. Their delicate flavor and texture has made them one of the most sought-after and expensive shellfish, and their classical "shell" shape has been used as an architectural motif for ages.

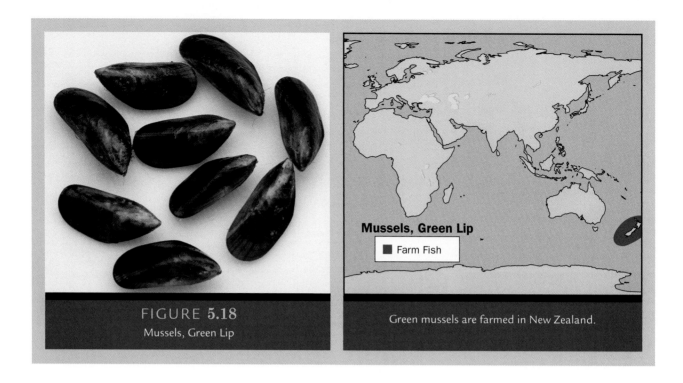

FIGURE 5.18
Mussels, Green Lip

Green mussels are farmed in New Zealand.

Unlike oysters and clams, scallops are migratory and avid swimmers; they have a large and developed adductor muscle that is 100 percent edible. This muscle includes the roe or coral, which is white in males and orange/red in females.

Because they cannot keep their shells closed, scallops lose moisture quickly and most are shucked soon after dredging or harvesting. Some sea scallops are collected by divers from vessels referred to as day boats. These are typically sold in their shell and command a premium.

SCALLOP, BAY (ARGOPECTEN IRRADIANS)

Once plentiful in bays and estuaries along the East Coast from Canada to North Carolina, the bay scallop has been depleted by overfishing, coastal development, and pollution. Current regulations limit landings; most of the bay scallops available in the United States are cultured on suspension nets in China. About the size of the tip of a pinky finger, bay scallops range in color from white to faint orange.

QUALITY CHARACTERISTICS

They should smell and taste sweet, nutty, and briny. Like sea scallops, bays are treated with sodium tripolyphosphate (STP) especially when sold frozen. Bay scallops should be cooked quickly or they will toughen and dry out. Deep frying, broiling, sautéing, and stir frying are all techniques that highlight their flavor and texture.

PREPARATION METHODS

- Bake
- Broil
- Fry
- Grill
- Poach
- Sauté
- Smoke
- Steam
- Stir fry

SEASONAL AVAILABILITY

Farm-raised bay scallops are available year-round, wild are available from October to May.

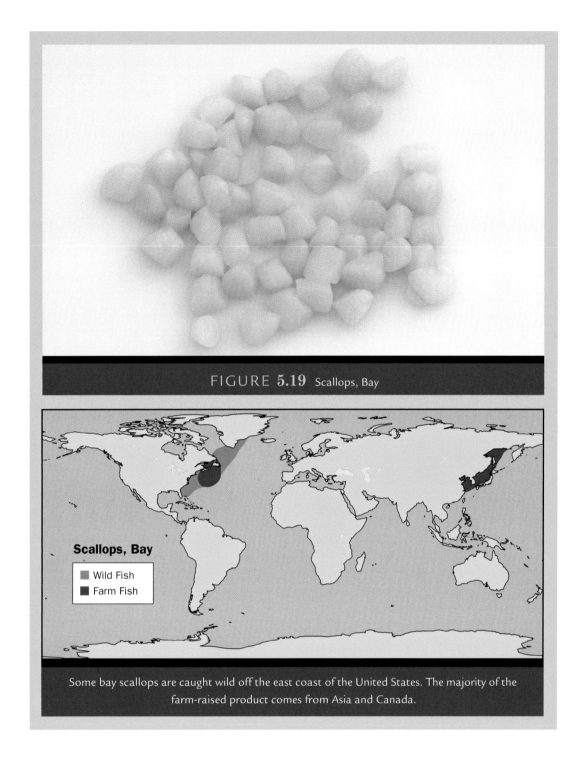

FIGURE 5.19 Scallops, Bay

Some bay scallops are caught wild off the east coast of the United States. The majority of the farm-raised product comes from Asia and Canada.

SCALLOP, BAY (3 OZ SERVING)

Water: 66.78 g	Calories: 75	Total fat: 0.65 g	Saturated fat: 0.067 g
Omega-3: 0.161 g	Cholesterol: 28 mg	Protein: 14.26 g	Sodium: 137 mg

SCALLOPS, SEA (*PLACOPECTEN MAGELLANICUS*)

Measuring up to 8 inches in diameter, sea scallops are the largest, most expensive, and abundant of all varieties, with concentrations off the coast of New England where they are dredged from depths of up to 800 feet. In Europe and Asia, they are sold with their roe or coral intact, but in the United States it is removed except on special order.

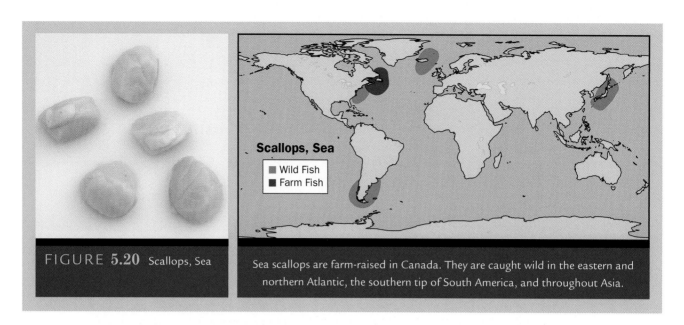

FIGURE 5.20 Scallops, Sea

Sea scallops are farm-raised in Canada. They are caught wild in the eastern and northern Atlantic, the southern tip of South America, and throughout Asia.

QUALITY CHARACTERISTICS

Scallops are sometimes soaked in sodium tripolyphosphate (STP), which enables them to absorb and retain moisture, adding to the price per pound, and making them nearly impossible to cook. Scallops that have been soaked tend to be very firm and white. Avoid these at all costs; insist on "dry pack" which will have more of a light flesh color and may be slightly "droopy" in appearance.

The flavor of sea scallops is sweet, slightly briny with a firm vertically layered texture; make sure they are very dry. Cook them using high-heat cooking methods such as searing or sautéing. Depending on their size, it may be beneficial to caramelize only one side to prevent overcooking. Classically, they are accompanied by delicate vegetables, butter sauces, and served in the shell when it is available.

PREPARATION METHODS

- Boil
- Broil
- Fry
- Pan Fry
- Sauté (picked meat or soft-shells)
- Steam
- Stir fry

SEASONAL AVAILABILITY

Year-round

SCALLOP, SEA (3 OZ SERVING)

Water: 66.78 g   Calories: 75   Total fat: 0.65 g   Saturated fat: 0.067 g
Omega-3: 0.161 g   Cholesterol: 28 mg   Protein: 14.26 g   Sodium: 137 mg

SEA URCHIN (*STRONGYLOCENTROTUS FRANSISCANUS, STRONGYLOCENTROTUS DROEBACHIENSIS*)

Eaten throughout the world and considered a delicacy, the sea urchin is a strange-looking marine animal resembling a pincushion. Known as *sea eggs* in the Caribbean

FIGURE 5.21 Sea Urchins

Sea urchins are harvested off of both coasts of the United States, and throughout the Caribbean, Mediterranean, and Asia.

and *oursins* in the Mediterranean, in the United States they are most typically served in Japanese restaurants where they are eaten raw or fermented and are called *uni*. With hundreds of species worldwide, these echinoderms are found on reefs, in tidal pools, and along the shore. Most popular is the green urchin, which is available on both coasts from late summer to spring. To harvest this delicacy, wear cut-resistant gloves. Hold the urchin with the hole or mouth side up. Using scissors, carefully cut around the mouth opening up the top side of the shell to expose the roe, which is firmly attached to the bottom part of the shell. Shake out the black viscera, and the roe is ready for use. It can also be purchased out of the shell from Japanese food purveyors.

QUALITY CHARACTERISTICS

The prize of these prickly round creatures is the roe or edible gonads, which is pale brown to orange in color. The female ovaries look like loosely compacted fish eggs, whereas the male gonads have a fine, soft grain-like texture and briny, slightly earthy flavor. In addition to raw, they can be emulsified into a sauce as one would make a beurre blanc. This rich, earthy sauce can then be poured back into the shell and eaten with crusty French bread and a crisp cold glass of white wine.

PREPARATION METHODS

- Bake
- Emulsify into sauces
- Ferment
- Make into compound butters
- Raw

SEASONAL AVAILABILITY

Year-round

SEA URCHIN (3.5 OZ SERVING)

Calories: 148   Total fat: 3.21 g
Cholesterol: 498 mg   Protein: 16 g

Source: Seafood Handbook

SHRIMP

Shrimp inhabit all the world's oceans and are abundant on both coasts of the United States. They are found in both warm and cold water and live in varying degrees of salinity. In the past several decades, advancements in aquaculture techniques and shipping have transformed the shrimp from a rarity to standard in the display cases of every corner supermarket. Most consumers are unaware of what they are purchasing and are unfortunately guided by price and size, not flavor. Like all shellfish, flavor is determined by species and environment, including diet, water temperature, and life cycle.

Wild-caught fresh shrimp are superior in flavor and texture to farm-raised. Many factors contribute to this difference, but as with any product, matching its natural habitat in a controlled environment such as a pond or farm is difficult, if not impossible, to do. Farm-raised shrimp have many advantages. They are frozen immediately, and are consistent in quality and reasonably priced. Because they are not fed prior to harvesting, their intestinal tract is usually very clean making them easier to work with. Unfortunately, because they live in such crowded conditions, water quality has to be controlled with chemicals and the shrimp do not get enough exercise. All these factors combined, contribute to textural and flavor differences when compared to fresh varieties. In the wild, shrimp eat a variety of food; in captivity, they are fed pellets of fish meal, grains, and often medication to control diseases common to the intensive aquaculture environment.

FRESH SHRIMP

Like many species of shellfish, shrimp are seasonal. They are caught by trawlers and are highly perishable; speed to market is essential. A fresh shrimp has a wonderful sweet flavor and texture that can be highlighted with a variety of seasonings and cooking methods. Freshness is critical, so deep water varieties are often frozen at sea. Otherwise, the shrimp should be packed in ice and eaten as soon as possible. Avoid those shrimp that have a strong iodine smell, a sign that they have been out of the water too long. Because they freeze well, frozen shrimp are a suitable alternative to fresh, however, fresh are always available, so insist on them from your purveyor or supermarket.

PREPARATION METHODS

- Bake
- Deep fry
- Poach
- Smoke
- Stir fry
- Broil
- Grill
- Sauté
- Steam

SEASONAL AVAILABILITY

Year-round for each variety of shrimp since the majority of the catch is frozen due to it being highly perishable.

SHRIMP VARIETIES

SHRIMP, BLACK TIGER (*PENAEUS MONODON*)

Along with Pacific whites, black tigers are the most commercially marketed shrimp in the United States. Produced throughout southern Asia in large aquaculture operations, they get their name from the dark gray to black stripes across their backs. The relatively low cost of the cultured product makes them one of the most available generic "shrimp" in the market. Black tigers are gray in color with characteristic tiger stripes and should not be spotted black, which can be a sign of melanosis.

QUALITY CHARACTERISTICS

Black tigers are firm in texture and easily overcooked. Their mild flavor varies depending on water conditions, feed, and environment. If water conditions are not in balance, the flavor can be muddy. Always rinse and dry these shrimp before cooking.

SHRIMP, TIGER (3 OZ SERVING)

Water: 64.48 g   Calories: 90   Total fat: 1.47 g   Saturated fat: 0.129 g
Omega-3: 0.268 g Cholesterol: 129 mg Protein: 17.26 g   Sodium: 126 mg

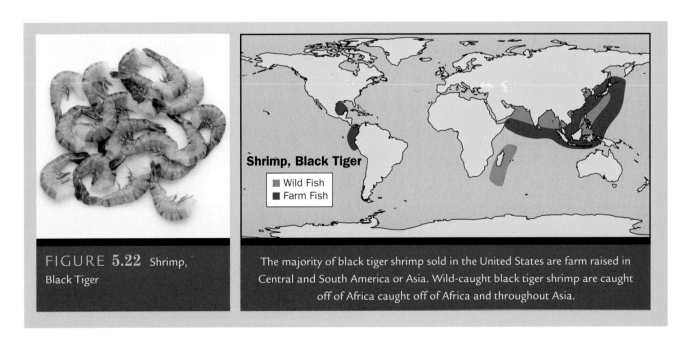

FIGURE 5.22 Shrimp, Black Tiger

The majority of black tiger shrimp sold in the United States are farm raised in Central and South America or Asia. Wild-caught black tiger shrimp are caught off of Africa caught off of Africa and throughout Asia.

SHRIMP, GULF (WARM WATER VARIETIES)

- Whites (*Penaeus setiferus*)
- Pinks   (*Penaeus duorarum*)
- Browns (*Penaeus aztecus*)

Gulf shrimp encompass a variety of different species that inhabit the Atlantic from Maryland south through the western Gulf of Mexico and are caught by shrimp trawlers. Identification can be difficult because colors vary depending on species and habitat. Of all the gulf varieties, the pink shrimp are the largest, growing to around 10 inches. Whites and browns range in size from 7 to 9 inches and all have their specific habitat concentrations, which account for a variety of different flavors.

QUALITY CHARACTERISTICS

Gulf shrimp are available year-round and have a mild but distinctive shrimp flavor. Texture is firm and they are slower to cook than their farmed cousins. Of the many varieties of Gulf shrimp, whites are said to have the best flavor. All Gulf shrimp will have a translucent appearance and be white, pink, or brownish gray in color. Gulf whites are typically whitish gray compared to the Pacific varieties which are more of a true white color. The shells of all varieties turn a light orange red-to-pink when cooked.

SHRIMP, GULF (BASED ON MIXED SPECIES DATA)

Water: 64.48 g
Calories: 90
Total fat: 1.47 g
Saturated fat: 0.129 g
Omega-3: 0.268 g

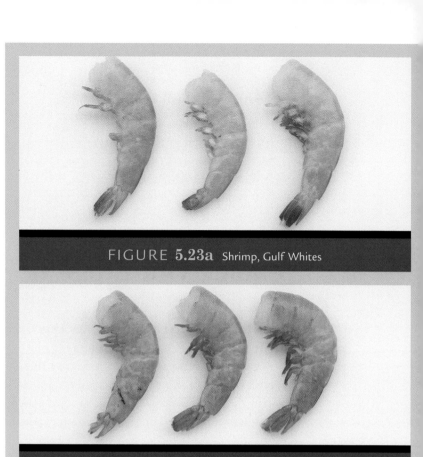

FIGURE 5.23a Shrimp, Gulf Whites

FIGURE 5.23b Shrimp, Gulf Pinks

FIGURE 5.23c Shrimp, Gulf Browns

Gulf shrimp, white, pink, and brown are harvested from the Carolinas south through the Gulf of Mexico.

Cholesterol: 129 mg
Protein: 17.26 g
Sodium: 126 mg

SHRIMP, PACIFIC WHITE (*PENAEUS VANNAMEI*)

Pacific whites share with black tigers the distinction of being the most farmed shrimp in the world. Raised throughout Asia, Central America, and South America, they can withstand very low levels of salinity and reach harvesting size in 3 to 4 months.

World markets have historically been concentrated in China and Ecuador with each country encountering difficulties in production. In the late 1990s, white spot disease decimated the Ecuadorian farms, and Chinese seafood tariff enforcement restricted the purchasing of international shrimp. To solve this import problem, the Chinese began raising the Pacific white shrimp throughout the country, realizing high profits from both the fresh and frozen market. This success lasted until 2007 when the European Union stopped shipments of Chinese white shrimp that contained the banned antibiotic chloramphenicol. Following suit, the state of Louisiana began testing imported shrimp for the antibiotic and lobbied the FDA to lower their accepted tolerance levels of chloramphenicol from 5 parts per billion to the European standards of 0.3 parts per billion.

Despite these sporadic setbacks, world supply of white shrimp continues to increase. Countries throughout Asia and Latin America are investing in aquaculture and are working on controlling disease and addressing problems with medications. As with any commodity, the price of shrimp is controlled by supply and demand. Current prices have seen historic lows especially for small and medium sizes, which continue to be the best value for the money. Prices for Mexican and U.S. white shrimp tend to be lowest at the end of the season in November when processors clear out their inventories in anticipation of colder temperatures.

QUALITY CHARACTERISTICS

Pacific white shrimp have many of the same qualities as their western counterparts. Because most of the Asian shrimp are farm-raised, often in low salinity water, the

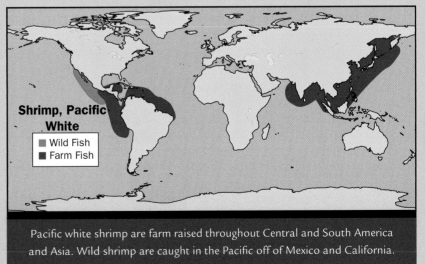

FIGURE 5.24 Shrimp, Pacific White. Pacific white shrimp are farm raised throughout Central and South America and Asia. Wild shrimp are caught in the Pacific off of Mexico and California.

flavor will not be as pronounced as wild ones or those farm-raised in ponds with higher percentages of salt water.

SHRIMP, PACIFIC WHITE (3 OZ SERVING)

Water: 64.48 g	Calories: 90	Total fat: 1.47 g	Saturated fat: 0.129 g
Omega-3: 0.268 g	Cholesterol: 129 mg	Protein: 17.26 g	Sodium: 126 mg

SHRIMP, PINK (COLD WATER VARIETY) (*PANDALUS* SPP.)

Pink shrimp are found in deep waters in most of the cold northern oceans of the world and are caught by trawling along muddy bottoms.

QUALITY CHARACTERISTICS

Mild in flavor, the pink shrimp have a sweeter flavor and noticeably softer texture than warm water shrimp. The tail meat is red to pink in color when raw and white with pink highlights when cooked. Pink shrimp are small, averaging about 35 per pound on the large size, to about 200 per pound for the smallest varieties. Many are processed and frozen at sea. Because of their small size they can easily overcook so high-heat cooking methods such as deep frying, steaming, and stir frying are preferred.

SHRIMP, PINK (3 OZ SERVING)

Water: 64.48 g	Calories: 90	Total fat: 1.47 g	Saturated fat: 0.129 g
Omega-3: 0.268 g	Cholesterol: 129 mg	Protein: 17.26 g	Sodium: 126 mg

SHRIMP, ROCK (*SICYONIA BREVIROSTRIS*)

Rock shrimp are crustaceans found in the Atlantic Ocean from Virginia to Mexico and average in size from 2 to 5 inches or 20 to 30 per pound. Named for their extremely hard shell, they are trawled throughout the summer and fall using reinforced nets. The majority of the catch is caught off of Florida's east coast.

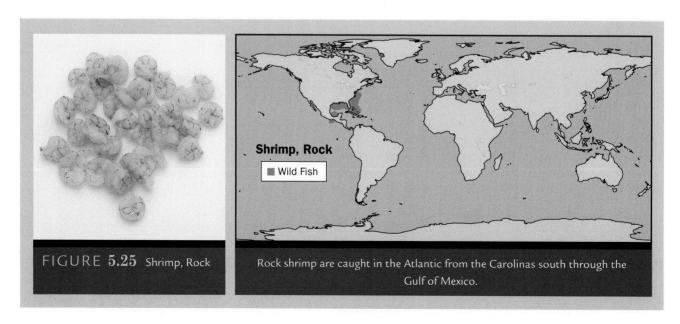

FIGURE 5.25 Shrimp, Rock

Rock shrimp are caught in the Atlantic from the Carolinas south through the Gulf of Mexico.

QUALITY CHARACTERISTICS

Firm in texture, and mildly sweet in flavor, the rock shrimp is similar to the spiny lobster. Due to their small size, they easily can be overcooked, making them dry and tough. They are available fresh, frozen, shell-on, or split and deveined and are often mislabeled on restaurant menus.

SHRIMP, ROCK (3 OZ SERVING)

Water: 64.48 g Calories: 90 Total fat: 1.47 g Saturated fat: 0.129 g
Omega-3: 0.268 g Cholesterol: 129 mg Protein: 17.26 g Sodium: 126 mg

SURIMI

Surimi is a processed form of fish that can be manipulated into a variety of shapes. Originating in Asia, it is made by pureeing washed lean fish such as pollock into a protein-rich paste to which flavoring, color, starch, and various preservatives are added. Surimi can also be used to bind actual fish and shellfish to give the appearance of a higher-end product. Popular forms of surimi include crab legs, lump crab, shredded crab, shrimp, lobster, a wide range of seafood burgers, and sausage. Because surimi is cooked and often pasteurized, its shelf life far exceeds that of fresh seafood.

QUALITY CHARACTERISTICS

Surimi is pure white in color and often contains a coating of food dye to simulate the natural color of cooked shellfish. Texture varies depending on the product and manufacturer but generally should mimic that of cooked shrimp. Avoid those products that contain abundant amounts of preservatives or have unusually soft or firm textures.

PREPARATION METHODS

Since surimi is a fully cooked product it is ready to eat or can be used in any recipe that calls for crab or shrimp.

FIGURE 5.26 Surimi

SURIMI (3.5 OZ SERVING)

Calories: 99 Total fat: 0.9 g Saturated fat: 0.2 g Omega-3: 0.4 g
Cholesterol: 30 mg Protein: 15.2 g Sodium: 143 mg

CEPHALOPOD AND OTHERS IDENTIFICATION

OCTOPUS (*OCTOPUS DOFLEINI, OCTOPUS VULGARIS*)

A cephalopod with eight arms and a globular body, the octopus is available in a variety of market sizes. Found in all the world oceans, it is especially prized throughout Spain, Portugal, Italy, Greece, North Africa, and Asia. Because of its unusual shape and appearance, it is not popular in the United States outside of ethnic markets, where it is available fresh, frozen, or dried. Octopi are caught by trawling or in baited vase-like traps or wicker baskets that have a narrow opening.

QUALITY CHARACTERISTICS

Octopus is firm and flavorful and is normally sold cleaned with the eye removed. The meat is prized in Japan where it is served sushi-style, dried, pickled, or smoked. Throughout Southern Europe and the Mediterranean, it is stewed or cooked in its own ink. Fabrication involves turning it inside out and removing the beak and guts. The skin of the octopus is edible and peels off easily when cooked.

PREPARATION METHODS

- Bake
- Broil
- Fry
- Grill
- Sauté
- Soups
- Stews

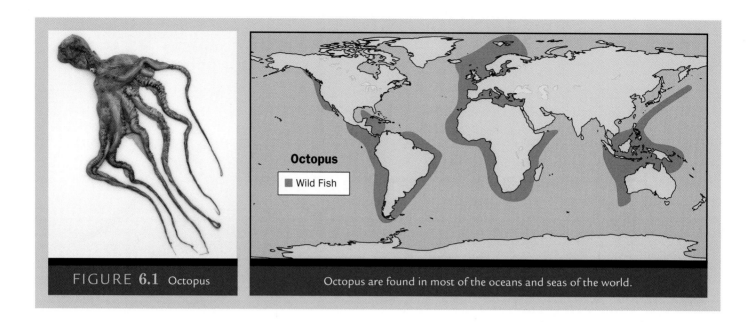

FIGURE 6.1 Octopus

Octopus are found in most of the oceans and seas of the world.

SEASONAL AVAILABILITY
Year-round fresh or frozen

OCTOPUS (3 OZ SERVING)

Water: 68.21 g	Calories: 70	Total fat: 0.88 g	Saturated fat: 0.193 g
Omega-3: 0.267 g	Cholesterol: 41 mg	Protein: 12.67 g	Sodium: 196 mg

OYSTERS

The term *oyster* encompasses a variety of bivalve mollusks found in both salt and brackish waters worldwide. Oysters are filter-feeders and have historically been credited with their ability to filter out large amounts of excess nutrients and pollutants from a variety of bays including the Chesapeake Bay.

Oysters inhabit depths of 12 to 30 feet, living in clusters and beds. Their shape often is determined by their habitat when they first begin to grow. Most wild oysters are gathered by hand or by rake. They can also be dredged, a practice that can damage the sea bottom. Dredging is banned or limited in many areas.

PREPARATION METHODS

- Can
- Ceviche
- Fry
- Pickle
- Raw on the half shell
- Smoke
- Steam
- Stew

OYSTER CULTIVATION

Oysters have been cultivated for many years. Small specimens called *spat* are seeded and allowed to mature on oyster beds or placed in submerged racks or bags until they reach harvest size in about a year.

OYSTERS AND THE MONTH WITH "R"

Whether it is an urban myth or partially true, many people believe that oysters should not be consumed in a month without the letter "R." There are perhaps two reasons for this belief. The first is that the months from May through August are the hottest of the year, making it difficult to maintain proper shipping temperatures. The other is that the oysters spawn during the warmer months and their meat becomes spongy, bland, and shrunken.

QUALITY CHARACTERISTICS

Oysters are a nutritionally sound food high in protein as well as vitamins A, B1, B2, B3, C, and D. They are also low in cholesterol and rich in zinc, phosphorus, calcium, magnesium, and iron. Although oysters can live up to two weeks out of water in a cold, damp environment, they are at their peak when freshly harvested.

Comparable in flavor complexity to wine, they can be sweet, briny, buttery, or taste subtly of minerals. Their texture should always be soft and slippery without being spongy and waterlogged, a sign that they have recently spawned.

Oysters can be prepared in many ways, but the consensus is that their flavor is most pronounced when consumed raw. All oysters eaten raw should be firmly closed and alive. Because they may contain harmful bacteria (mostly from unregulated waters), pregnant woman or those with compromised immune systems should avoid consuming any raw fish or shellfish.

OYSTER VARIETIES

Of the many varieties of oysters worldwide, there are five species of culinary importance, within which are many shapes and flavor differences depending on environment, salinity, and water quality. Many feel that oysters from colder waters have a greater depth of flavor and texture.

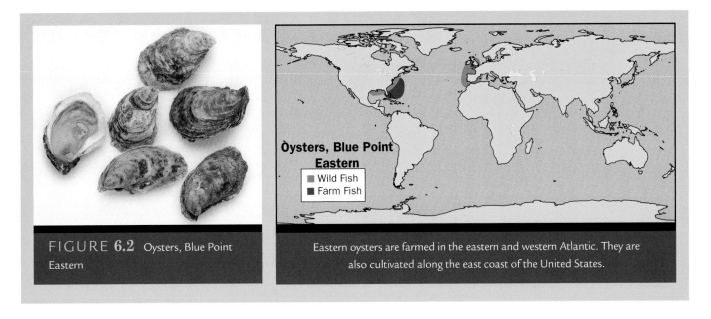

FIGURE 6.2 Oysters, Blue Point Eastern

Eastern oysters are farmed in the eastern and western Atlantic. They are also cultivated along the east coast of the United States.

OYSTER, EASTERN (*CRASSOSTREA VIRGINICA*)

This oyster is perhaps the most common in the United States. It measures 2 to 5 inches across and can be found from the Gulf of Mexico to Prince Edward Island.

REGIONAL NAMES

- Apalachicola (Florida)
- Bluepoint (New York)
- Breton (Louisiana)
- Chincoteague (Virginia)
- Malpeque (Canada)
- Wellfleet (Cape Cod)

PREPARATION METHODS

- Can
- Fry
- Raw on the half shell
- Smoke
- Steam
- Stew

SEASONAL AVAILABILITY

Available year-round with the warmer months being less desirable

OYSTER, EASTERN (3.5 OZ SERVING)

Calories: 137 Total fat: 4.9 g Saturated fat: 1.5 g Omega-3: 1.2 g
Cholesterol: 105 mg Protein: 14.1 g Sodium: 422 mg

Source: Seafood Handbook, their source USDA

OYSTER, EUROPEAN (*OSTREA EDULIS*)

A native of Brittany, France, the Belon can be distinguished by its large, round size and flat shell. Eaten raw on the half shell, it is prized for its remarkable flavor and texture. It should not be cooked.

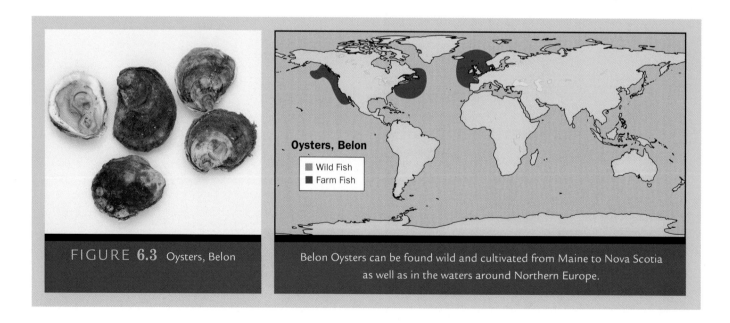

FIGURE 6.3 Oysters, Belon

Belon Oysters can be found wild and cultivated from Maine to Nova Scotia as well as in the waters around Northern Europe.

REGIONAL NAMES

- Blue hill (Maine)
- Belon (France and United States)
- Dorset (England)

OYSTER, EUROPEAN (3 OZ SERVING)

Water: 69.75 g	Calories: 69	Total fat: 1.95 g	Saturated fat: 0.433 g
Omega-3: 1.170 g	Cholesterol: 42 mg	Protein: 8.03 g	Sodium: 90 mg

OYSTER, KUMOMOTO (*CRASSOSTREA SIKAMEA*)

This oyster gets its name from the Japanese prefecture where it was first cultivated. Currently, it is a popular species in the Pacific Northwest where it has been cultivated for years. Rich and meaty in flavor, it is briny and very fragrant.

OYSTER, OLYMPIA (*OSTREA LURIDA*)

Native to the Pacific Northwest, this small species measures up to 2 inches and has a strong flavor sometimes associated with minerals.

OYSTER, PACIFIC (*CRASSOSTREA GIGAS*)

One of the largest varieties, the Pacific oyster is very fast growing and can reach up to 12 inches in length. Because of its rapid growth, it is widely cultivated and is harvested from Mexico to southern Alaska. Its size limits its value on the half shell, and the larger varieties can be tough. Its flavor is sweet and rich with characteristics of underripe melon.

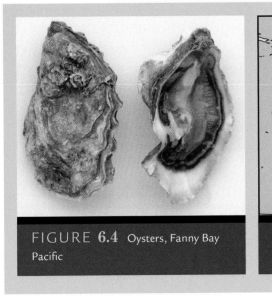

FIGURE 6.4 Oysters, Fanny Bay Pacific

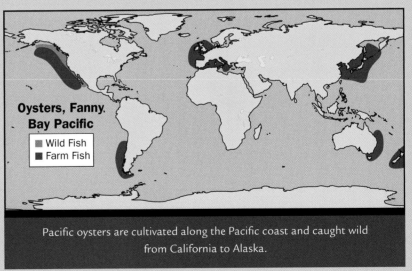

Pacific oysters are cultivated along the Pacific coast and caught wild from California to Alaska.

OYSTER VARIETIES

REGIONAL NAMES

- Japanese oyster
- Mad river (California)
- Totten (Washington State)

OYSTERS, PACIFIC, INCLUDING ALL THE ABOVE WESTERN VARIETIES (3 OZ SERVING)

Water: 69.75 g	Calories: 69	Total fat: 1.95 g	Saturated fat: 0.433 g
Omega-3: 1.170 g	Cholesterol: 42 mg	Protein: 8.03 g	Sodium: 90 mg

SQUID (*LOLIGO ILLECEBROSUS, LOLIGO OPALESCENS, LOLIGO PEALEI*)

Squid inhabit all of the world's oceans and are caught at night using bright attracting lights that coax them to the surface in search of prey. Often they can be pumped onboard with large suction-like devices, or by purse seines and trawlers. Squid are high in protein, low in fat, and are at least 75 percent edible, a high percentage for any type of seafood. Their skin is speckled gray with shades of purple, and white when cleaned. Sizes vary depending on location but generally run in the 3 to 6 inch range.

QUALITY CHARACTERISTICS

Several species of squid are caught and marketed in the United States. From a quality standpoint, the Pacific *Loligo opalescens* is the best. When purchasing fresh squid they should be moist, almost slimy, and should be washed before fabricating. Squid has tender flesh and a sweet, mild flavor. All squid start out tender, but get tough when overcooked. The rule for squid is to cook it 45 seconds or 45 minutes, that means quick cooking methods or moist heat such as braising and stewing. Squid are sold in many market forms including fresh, cleaned, rings, and tubes. Avoid frozen squid that are unnaturally white; these have been soaked in a brine solution which enhances color but not flavor. Squid ink is also available frozen and can be found in Asian markets or specialty purveyors.

PREPARATION METHODS

- Broil
- Fry
- Sauté
- Dry
- Poach
- Stew

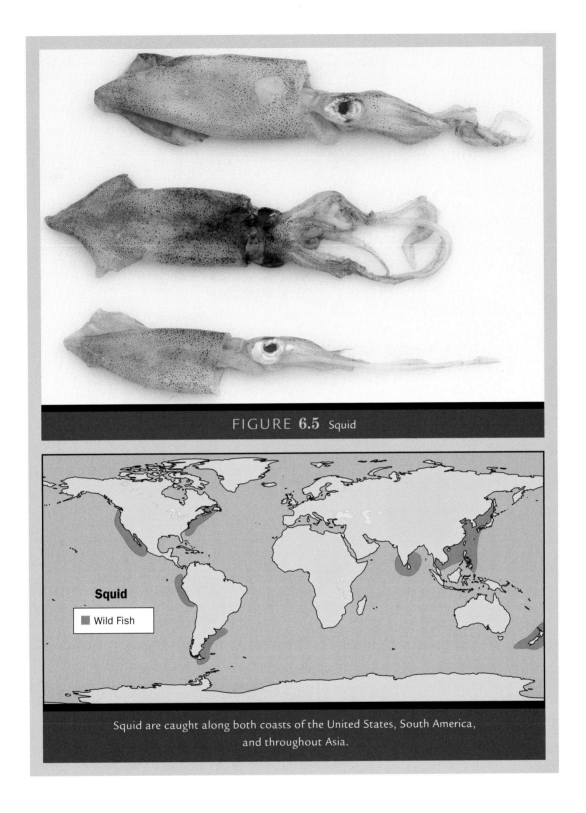

FIGURE 6.5 Squid

Squid are caught along both coasts of the United States, South America, and throughout Asia.

SEASONAL AVAILABILITY

Year-round

SQUID (3 OZ SERVING)

Water: 66.77 g   Calories: 78   Total fat: 1.17 g   Saturated fat: 0.304 g
Omega-3: 0.68 g Cholesterol: 198 mg   Protein: 13.24 g   Sodium: 37 mg

FIN FISH FABRICATION

FABRICATION

Much of the fish purchased by chefs and home cooks is filleted by hand or with specialized filleting machines. These fillets have become so commonplace that whole fish are often more difficult and expensive for the average multi-product commercial food seller to obtain and transport to the local restaurant or retail level operation. These companies push the pre-packaged fillets on chefs with the idea that fabrication is too cumbersome, time consuming, and messy to undertake. When purchasing seafood for a restaurant, wholesale seafood sellers are much better equipped to handle whole fish and understand its quality than the large multi-product vendors. The reality is that filleting fish is a messy process that takes time, space and skill, but freshness indicators such as clear eyes, aroma, bright gills, and firmness of the flesh cannot be evaluated if the fish is not whole.

Along with evaluation of the quality and freshness, it is much easier to determine the species by looking at a whole fish as opposed to fillets, with the added advantage that the leftover bones can be used to make valuable fish stock.

FISH GUTTING, SCALING, AND FABRICATION

GUTTING

Because the enzymes in the viscera (guts) rapidly break down the flesh leading to spoilage, it is extremely important to remove them as soon as possible after the fish is taken from the water. To gut a round fish, place the tip of the knife in the anal opening close to the tail, and carefully cut up to where the gills meet the head (see Fig 7.1a). Do not insert the knife too deeply into the cavity or the internal organs will be pierced, causing a foul-smelling mess.

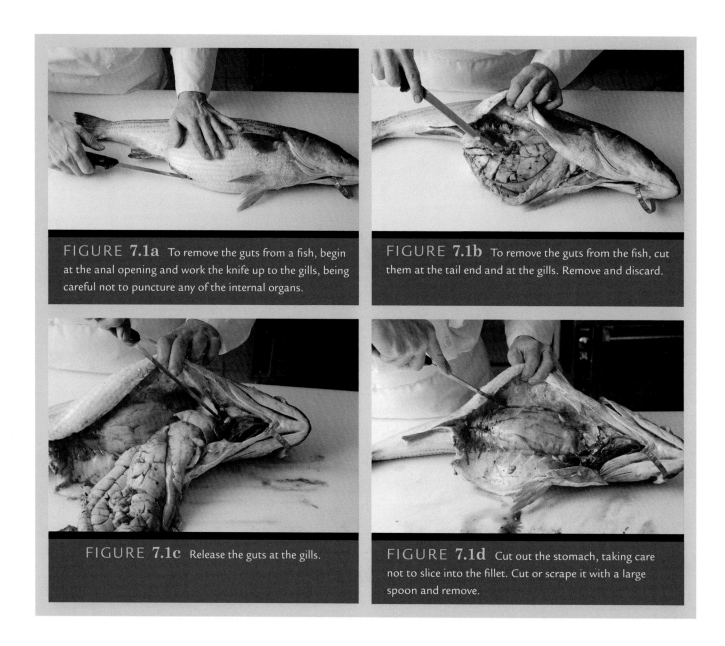

FIGURE **7.1a** To remove the guts from a fish, begin at the anal opening and work the knife up to the gills, being careful not to puncture any of the internal organs.

FIGURE **7.1b** To remove the guts from the fish, cut them at the tail end and at the gills. Remove and discard.

FIGURE **7.1c** Release the guts at the gills.

FIGURE **7.1d** Cut out the stomach, taking care not to slice into the fillet. Cut or scrape it with a large spoon and remove.

Once the cavity has been opened and the internal organs are visible, make a cut behind the head and next to the tail to separate the organs from the body (see Fig 7.1b). Remove and discard the viscera; if the head is to be removed, do it at this point (see Fig 7.1c).

Examine the interior cavity and with a small spoon (see Fig 7.1d), or a pulling motion, take out the dark vein that runs along the backbone and any viscera that may have been missed. Then give the cavity a final thorough rinsing.

SCALING

Nearly all fish have scales that must be removed prior to fabrication. Scales reveal a lot about a fish, but for the most part they exist for external protection and must be removed. The size, shape, and type of scales vary from species to species; they can be very small

FIGURE 7.2 Scale fish by running the scaling tool from the tail to the head in a firm consistent motion.

or large enough to be used for jewelry. There are many types of scalers available on the market, some made out of hard plastic and others metal. They are designed to remove the scales without damaging the skin by drawing the tool against the scales in a brushing motion that snaps them off. The back of a fillet knife, a common table knife, or a spoon can also be used but not as effectively.

Place the fish in a perforated hotel pan under running water to contain the scales. Hold the fish by the tail and work the scaling tool from the tail to the head with enough pressure to remove the scales without bruising the flesh (see Fig 7.2). Repeat this process on the reverse side until all the scales have been removed and the fish is rinsed clean.

FABRICATION OF FLAT FISH AND ROUND FISH

From the standpoint of fabrication, most fish fall into two broad categories, round and flat. The methods used for removing the fillets are consistent within these two groupings. There are certainly exceptions like the skate (see page 87) but those will be dealt with on an individual basis.

Flat fish fabrication

The body shape of flounder, sole, and halibut is oval and compressed. These fish live relaxed lives on the bottom, giving their meat a delicate texture and flavor. Flat fish have one large fillet on either side of their body that can easily be broken down into two or more separate pieces. Even a 700 pound halibut is fabricated the same way because its body and bone structure is similar to the smallest flounder.

The fillets on all flat fish are located on either side of the body, which is dark on one side and light on the other. The dark skin side tends to be slightly thicker so when filleting larger fish, place that side down on the cutting board and begin your first cut on the lighter side. The thicker meat will support the weight better and ensure that the fillets are not crushed. Whether you remove the whole side or just half will depend on the size of the fish and the application, but remember, larger pieces store and cook better than smaller ones.

MISE EN PLACE

With all kitchen operations that require any kind of setup, organization is paramount to success. In French this concept is referred to as *mise en place*, and it allows for a logical workflow, including station setup and cleanup. Proper mise en place for fabrication includes having all the proper knives, tools, and equipment set out and ready for use. Water is an important element in fabricating fish. It washes away viscera, scales, slime, and blood, enhancing the final product. Sharp knives also make the job easier as does the proper selection of pans and tubs to store the fabricated fish.

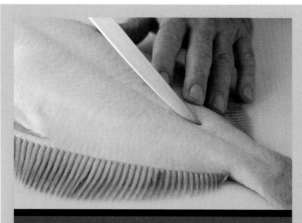

FIGURE **7.3a** The lateral line running up the middle of all flat fish is a good indicator of where to begin cutting when removing half fillets.

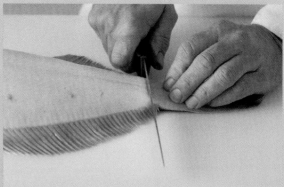

FIGURE **7.3b** Cut across the tail, and down to the bone.

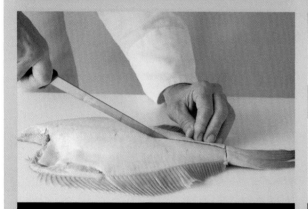

FIGURE **7.3c** Following the backbone and lateral line, cut down to the bone, from the tail to head.

FIGURE **7.3d** Using a flexible fillet knife, start at the tail and slowly separate the fillet from the bones. Use the tip of the knife and keep it flat on the bones as you draw the knife toward the head of the fish until the fillet is removed.

FIGURE **7.3e** After the first fillet has been removed, turn the fish around and repeat the procedure, removing the fillet from the head to the tail.

FIGURE **7.3f** Using a flexible fillet knife, start at the tail and slowly separate the fillet from the bones.

(continues)

(continued)

FIGURE **7.3g** Use the tip of the knife and keep it flat on the bones as you draw the knife toward the head of the fish until the fillet is removed.

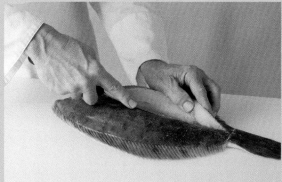

FIGURE **7.3h** Turn the fish over and repeat the procedure to remove two more fillets.

FIGURE **7.3i** Trim the sides of the fillets and save for fish stock.

FIGURE **7.3j** To skin the fillets, keep the knife flat while working it under the meat at the tail end. Pull the skin toward you while pushing and sawing with the knife.

FIGURE **7.4a** To remove the entire fillet from a flat fish, place one hand on top of the fish for stability and begin cutting on top of the fish near the head. Keep the knife flat on the bones and slowly cut with a sawing or swiping motion until the knife reaches the center backbone.

FIGURE **7.4b** Once the knife has reached the center backbone, go up and over the bone and work the tip of the knife down the bones of the other side.

(continues)

FABRICATION **163**

(continued)

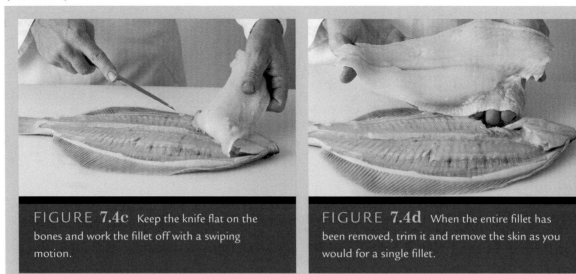

FIGURE 7.4c Keep the knife flat on the bones and work the fillet off with a swiping motion.

FIGURE 7.4d When the entire fillet has been removed, trim it and remove the skin as you would for a single fillet.

FIGURE 7.5 One way to cook a flat fish fillet is to roll it into a paupiette. To do this, roll the fillet with the skin side up so the belly side (presentation side) is facing up.

FIGURE 7.6 Another form a flat fish fillet can take to be cooked is to fold the fillet in half or thirds creating a compact portion for poaching.

Unlike round fish that have many small bones throughout their bodies, flat fish can easily be filleted totally boneless. When combined with onions, carrots, celery, and water these bones make a superior fish stock. The skin is typically not eaten and can be added to the stock.

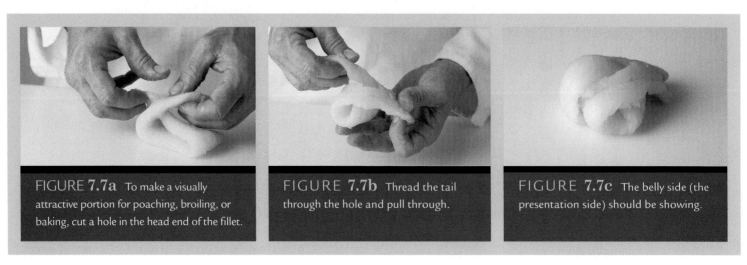

FIGURE 7.7a To make a visually attractive portion for poaching, broiling, or baking, cut a hole in the head end of the fillet.

FIGURE 7.7b Thread the tail through the hole and pull through.

FIGURE 7.7c The belly side (the presentation side) should be showing.

164 CHAPTER 7 · FIN FISH FABRICATION

Round fish fabrication

The fusiform-shaped bodies of the round fish lend themselves to many styles of fabrication depending on the size and intended use. Smaller round fish like bass and salmon have two distinct fillets on either side of their body. Larger specimens like the 1000-pound. bluefin tuna can be broken down in four or more distinct sections each having its own unique qualities. A wide variety of fin and bone structures make many of them slightly different to deal with, but as a group all round fish have the same characteristics.

FILLETING A LARGE ROUND FISH

To remove the two fillets from a fish like a salmon, lay the fish on a cutting board with the head on your knife hand side. Make a cut through the belly flap behind the pectoral fin and head, angling the knife toward the head as you cut down to the bone (see Fig 7.8a). Because it has the ability to extend out through the fish, a long fillet or slicing knife works well for this type of fish. Once the first cut is complete, snap your wrist around so the knife is laying flat on the backbone with the edge facing the tail of the fish (see Fig 7.8b). At this point, hold the upper belly portion close to the pectoral fin with your opposite hand and draw the knife down the fish toward the tail, keeping pressure on the knife so it does not ride up into the fillet, or down through the backbone (see Fig 7.8c). While making this cut there will be some resistance as the knife cuts through some of the delicate internal bones. Once the cut is complete, the fillet can easily be removed. Next, turn the fish over so the tail is on the side of your knife hand and the backbone is facing you. Make your next cut at the base of the tail with an initial sawing motion until the knife begins to separate the second fillet from the bones. Hold the belly section up as you draw the knife down toward the head while keeping pressure on the knife as it cuts through the bones and releases the fillet. Once you have reached the head, separate the fillet by cutting the same angle cut behind the head and pectoral fin. With this method, there will be some meat left on the backbone, but this can easily be scraped off with a spoon

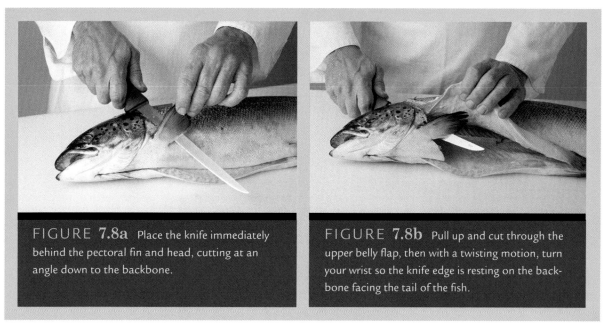

FIGURE **7.8a** Place the knife immediately behind the pectoral fin and head, cutting at an angle down to the backbone.

FIGURE **7.8b** Pull up and cut through the upper belly flap, then with a twisting motion, turn your wrist so the knife edge is resting on the backbone facing the tail of the fish.

(continues)

(continued)

FIGURE **7.8c** Keep the knife flat on the backbone and cut away from the head toward the tail using a sawing and pulling motion while cutting through the pin bones. Steady the fish with the opposite hand throughout the procedure.

FIGURE **7.8d** Using this method for filleting large round fish will necessitate using a spoon to scrape the bones for the several ounces of beautiful meat left behind. This meat has a wonderful flavor and texture and is extremely versatile and sought after for sashimi. Also it can be made into delicate mousseline quenelles.

FIGURE **7.8e** Trim the belly bones off of the fillet using a flexible fillet knife.

FIGURE **7.8f** To pull out the delicate pin bones, grasp them firmly with small pliers and pull toward the head.

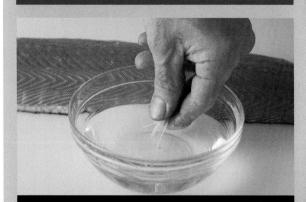

FIGURE **7.8g** Keep a bowl of water handy to loosen and collect the bones from your fingers.

FIGURE **7.8h** To remove the skin from the fillet, separate it from the tail and make a small hole in the tail end.

(continues)

(*continued*)

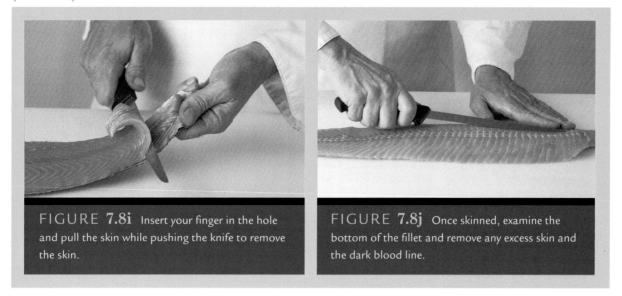

FIGURE 7.8i Insert your finger in the hole and pull the skin while pushing the knife to remove the skin.

FIGURE 7.8j Once skinned, examine the bottom of the fillet and remove any excess skin and the dark blood line.

and used for a myriad of preparations or frozen for future use (see Fig 7.8d). The fillets should be trimmed by cutting off the belly bones and flaps with a ridged knife (see Fig 7.8e), and any internal bones can be removed with tweezers or needle-nose pliers (see Fig 7.8f). To remove the skin, use the same fillet knife and work it under the meat and above the skin (see Fig 7.8h). Tug on the skin while gently pushing the knife toward the head end to remove the fillet. Turn the fillet over and remove any of the excess blood line with the fillet knife (see Fig 7.8j).

FILLETING A SMALL ROUND FISH

Although there is not much difference between a large and small round fish, the smaller fish, such as snapper and bass, are often filleted with a smaller knife that cuts the fillet off of the bones instead of cutting through the bones as in the description for large round fish. To fillet a smaller round fish, place the fish with the backbone toward you and the head at your knife hand. Make a cut behind the pectoral fin and head, angling the knife toward the head as you cut down to the bone (see Fig 7.11a). Make sure that the knife cuts through the belly flap while making this cut. Next feel for the backbone and make a cut along the backbone toward the tail as if you were cutting a long piece of French bread. Hold the fish steady with your other hand while you make this cut. Go slowly and use the tip of the knife, which should be kept flat

FIGURE 7.9 Cut tranches by slicing through the fillet at an angle.

FIGURE 7.10 For more of a rectangle fillet, slice straight down.

FIGURE **7.11a** To fabricate a small round fish, make the first cut at an angle behind the head and the pectoral fin, cut down to the backbone.

FIGURE **7.11b** Placing your other hand on the top fillet for support, begin to cut down the backbone toward the tail, keeping the tip of the knife firmly on top of the bone. Carefully remove the fillet from the bones until you reach the center bone.

FIGURE **7.11c** As the fillet is being separated from the belly bones, lift the fillet with the other hand to view the progress. Once at the center bone, go up and over the bone to the other side to remove the fillet. Be sure to put pressure on the knife to keep it straight and avoid cutting into the fillet.

FIGURE **7.11d** To remove the opposite fillet, turn the fish over so the tail is toward your knife hand and make a similar cut behind the head and the pectoral fin.

FIGURE **7.11e** Working from the tail, this time cut along the backbone using your other hand to support the fish and guide the knife which should be held firmly.

FIGURE **7.11f** Again, once at the center bone, go up and over the bone to the other side to remove the fillet. Be sure to put pressure on the knife to keep it straight and avoid cutting into the fillet.

on the bones (see Fig 7.11b). Continue removing the fillet until the tip of the knife hits the backbone. At this point the knife tip should go up and over the bone to the other side (see Fig 7.11c). Begin removing the belly portion of the fillet, drawing the tip of the knife down toward the tail. When the first fillet is removed, turn the fish over so the tail is at your knife hand and the backbone is facing you. Begin to fillet the fish from the tail side using the same technique as used on the first fillet (see Figs 7.11d through 7.11f).

REMOVING THE BONES FROM THE BELLY
Round fish are often stuffed or butterflied whole. To accomplish this, all the bones are removed from the belly in one piece. First make sure that the belly is cut open all the way to the tail, then locate the front rib bones located in back of the pectoral fins and behind the head. Insert a long, flexible fillet knife between the bones and the fillet and separate the meat from the rib bones from front to back (see Fig 7.12a). Once all of the rib bones are free on both sides of the fillet, lay the fish on its back and run the knife along the backbone removing it in one piece (see Fig 7.12b). Inspect the fish for bones and remove them as necessary (see Fig 7.12c). Typically a whole boneless fish is poached whole, stuffed, roasted, or cut into boneless steaks (see Fig 7.12e).

FIGURE **7.12a** Take notice of the position of the knife to separate the backbone and upper rib bones from the fillets.

FIGURE **7.12b** To bone out the fish from the belly, insert a long flexible fillet knife along the backbone and upper rib bones and separate them from the two fillets.

FIGURE **7.12c** Completely remove the backbone from the fish using a flexible fillet knife.

FIGURE **7.12d** Check the fillet for any pin bones and remove them using small pliers.

(*continues*)

(continued)

FIGURE **7.12e** After removing the belly bones and fins, cut through the upper fillet and slice through the backbone into and through the lower fillet using a chef's knife or ridged fillet knife.

FIGURE **7.12f** For consistent cooking and presentation, tie the steaks into rounds using butcher's twine.

AN EXCEPTION TO FLAT FISH FABRICATION: DOVER SOLE

FIGURE **7.13a** To prepare Dover sole for sautéing, first trim the fins off of the top and bottom with a pair of scissors.

FIGURE **7.13b** To remove the skin from Dover sole, make a cut at the tail and separate the skin from the meat.

FIGURE **7.13c** Grasp the skin at the base of the tail and pull it off toward the head in one piece. It is helpful to dip your fingers in salt to get a good grip on the skin.

FIGURE **7.13d** Repeat the procedure on the other side.

(continues)

(continued)

FIGURE **7.13e** The whole Dover sole has been gutted, trimmed, and skinned and is now ready for cooking.

The way in which Dover sole is fabricated is different from any other fish. Although a flat fish, Dover sole has unique characteristics that make its fabrication rather quick and easy with little need for a knife; all you really need is a pair of scissors and your hands.

SHELLFISH FABRICATION AND TOOLS OF THE TRADE

SHELLFISH FABRICATION

Despite the high labor cost involved, much of the shellfish we use arrives fabricated. Shrimp, crab, squid, and even clams are neatly cleaned and arranged in containers ready to use. A huge variety of market forms are available to meet any need and price range. Because shellfish is so perishable and does not travel well nor easily withstand temperature changes, it is often immediately cooked or frozen. Shrimp are de-headed, squid are cleaned and gutted, and crab is skillfully removed from the shell. These market forms, although convenient, not only impact the product but often come with a price that must be passed on to the customer.

There are many advantages to using whole fish and shellfish in their original form. Quality of the product can be easily determined when it arrives intact; often it is cheaper. However, depending on labor cost and your ability to utilize shellfish trim, it might be cost effective to purchase it fabricated. Deciding what to purchase also depends on location. Those living near the sea are much more likely to receive fresh unfabricated product than those thousands of miles inland.

Evaluating these factors will allow you to make logical choices when purchasing shellfish. Ultimately, quality and price should be the deciding factors in determining specific product forms.

LOBSTER FABRICATION

A chef's repertoire normally includes several lobster dishes that require the cooked meat to be removed from the shell as intact as possible. Pound for pound, lobster is among the most expensive items, so yield and proper cooking are important. Raw lobster meat is wet and jelly-like and is easier to work with when it is at least partially cooked prior to being removed from the shell. Some chefs quickly blanch the whole lobster to facilitate extraction. This process is tricky; because raw lobster has a short shelf life, blanching should be done immediately before final cooking. The other option is cooking it to an internal temperature of 140° to 145°F/60° to 63°C (see Fig 8.1) and removing the meat, a method that is safer as long as overcooking is avoided. With practice, removing the cooked meat from the shell is cost effective regardless of the time it takes, and is especially worth it when the shells are utilized for stock.

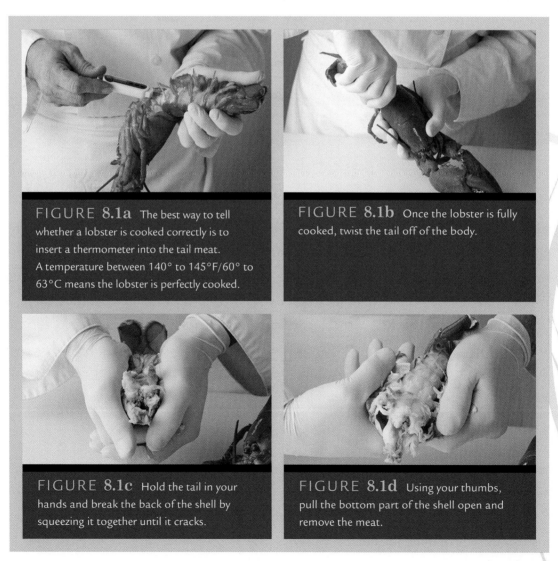

FIGURE 8.1a The best way to tell whether a lobster is cooked correctly is to insert a thermometer into the tail meat. A temperature between 140° to 145°F/60° to 63°C means the lobster is perfectly cooked.

FIGURE 8.1b Once the lobster is fully cooked, twist the tail off of the body.

FIGURE 8.1c Hold the tail in your hands and break the back of the shell by squeezing it together until it cracks.

FIGURE 8.1d Using your thumbs, pull the bottom part of the shell open and remove the meat.

(continues)

(continued)

FIGURE 8.1e To remove the claw and knuckle, twist them off of the body.

FIGURE 8.1f Remove the knuckle from the claw by snapping it off against the joint.

FIGURE 8.1g Separate the top shell from the body. Remove and discard the feathery gills. Pull off and reserve the legs. The body and upper shell can be made into stock.

FIGURE 8.1h To remove the claw meat, work off the pincher by carefully pulling it side to side and backward against the joint until it releases. (Note: Sometimes the inner cartilage will come out when you remove the pincher, but often it must be removed after the claw has been extracted from the shell.)

FIGURE 8.1i Using the heel of the knife, chop through the claw about 1/2 inch and quickly twist the knife to remove the bottom portion of the shell; the claw meat should easily pop out.

FIGURE 8.1j If the plastic-like cartilage is still in the claw, gently pull it out. A paring knife can also be used.

(continues)

SHELLFISH FABRICATION

(continued)

FIGURE 8.1k To remove the knuckle meat, cut the knuckle in half with scissors.

FIGURE 8.1l To remove the leg meat, line them up on a board and push the meat out with a rolling pin or the back of a chef's knife. Save the leg shells for stock.

FIGURE 8.1m Average yield from a 1-1/2 pound lobster.

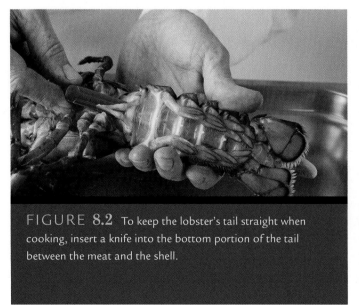

FIGURE 8.2 To keep the lobster's tail straight when cooking, insert a knife into the bottom portion of the tail between the meat and the shell.

FIGURE 8.3 The lobster tail on the left has been cooked, with the tail out straight, held in place with a common table knife that is left in during cooking. The tail on the right has naturally curled during cooking.

176 CHAPTER 8 · SHELLFISH FABRICATION

FIGURE 8.4 To humanely kill a lobster, insert the knife through the head between the eyes.

Another way to fabricate a lobster for cooking is to cut it in half while it is alive. This can be done humanely by first severing the brain, which is located behind the eyes. Once in half, the lobster can be seasoned, broiled, grilled, or roasted.

The tail meat of the lobster is the firmest, followed by the claws, knuckles, and feet. Inside the body is the green tomalley, which is the liver, and in females a bright red roe. Both are edible but many feel the liver can accumulate toxins.

SQUID FABRICATION

Squid are sold in a variety of forms; the most economical, depending on your labor cost and location (because they have a limited shelf life), are whole and must be cleaned. Along the East and West coast of the United States, large amounts of squid are trawled year-round and are available fresh at very reasonable prices. Squid freeze well, an advantage to people living inland. Avoid those products that have been whitened through brining because any unnatural processing can diminish consistency and flavor.

CLEANING SQUID

To clean fresh squid, pull the head (see Fig 8.5a) and tentacles away from the mantle (tube) and remove the inner plastic-like bones (quills). Cut the head from the tentacles just below the eyes; discard the head and save the tentacles. Scrape the dark skin off of the mantle with a knife (see Fig 8.5c). If the tubes are to be used whole for stuffing, rinse them well in cold water. Rings should be cut first and then washed along with the tentacles (see Fig 8.5d and 8.5e). Squid have an ink sac located below the head, which can be removed and passed through a sieve to collect the edible ink. Ink is also available from specialty purveyors frozen in small packages or 1/2-liter containers.

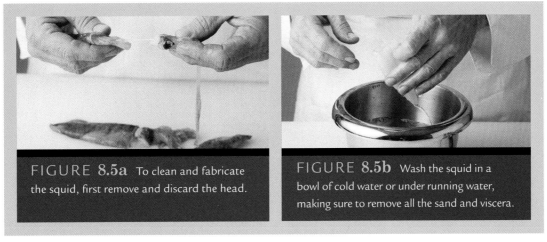

FIGURE 8.5a To clean and fabricate the squid, first remove and discard the head.

FIGURE 8.5b Wash the squid in a bowl of cold water or under running water, making sure to remove all the sand and viscera.

(*continues*)

(continued)

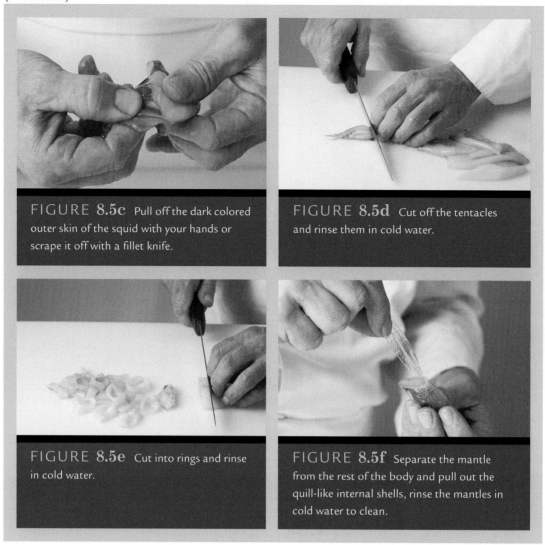

FIGURE 8.5c Pull off the dark colored outer skin of the squid with your hands or scrape it off with a fillet knife.

FIGURE 8.5d Cut off the tentacles and rinse them in cold water.

FIGURE 8.5e Cut into rings and rinse in cold water.

FIGURE 8.5f Separate the mantle from the rest of the body and pull out the quill-like internal shells, rinse the mantles in cold water to clean.

SCALLOP FABRICATION

If you are lucky enough to have access to scallops in the shell, you will notice that they are easier to open than most shellfish. These large bivalve mollusks are attached to a pair of shells that can be washed and used to serve French classic Coquilles St. Jacques.

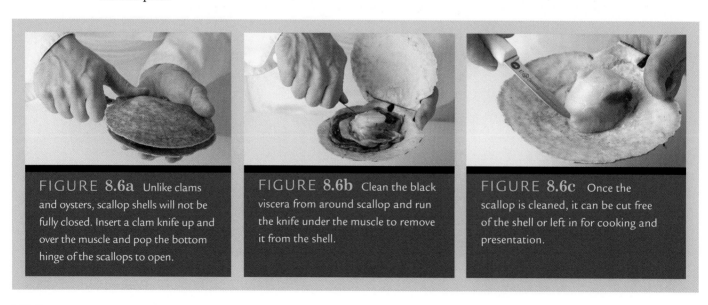

FIGURE 8.6a Unlike clams and oysters, scallop shells will not be fully closed. Insert a clam knife up and over the muscle and pop the bottom hinge of the scallops to open.

FIGURE 8.6b Clean the black viscera from around scallop and run the knife under the muscle to remove it from the shell.

FIGURE 8.6c Once the scallop is cleaned, it can be cut free of the shell or left in for cooking and presentation.

CRAB FABRICATION

The crab's hard and ridged shells contain meat that is almost jelly-like when raw, but lumpy and tender when cooked. It is typically steamed or boiled before the various crevasses and channels of the body and legs yield their delicious meat. The texture of the crab's meat depends on its size and species. Large legs from the king crab are split in half with saws in the processing plant for easier access to its dense but delicate meat with its spun string texture. Blue crabs are known for their body meat, which is skillfully removed and marketed as jumbo. Other species are available fresh cooked, frozen, or pasteurized. Fabrication is usually slow and tedious at first but is worth the effort and depending on your labor cost, worth the price.

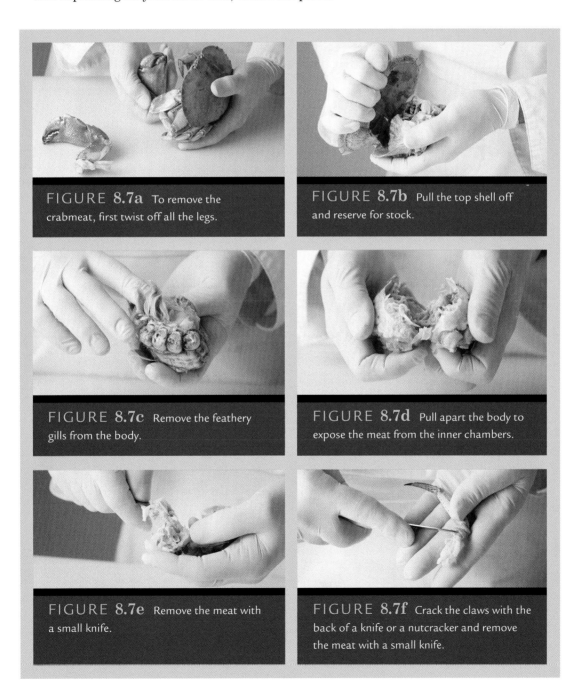

FIGURE 8.7a To remove the crabmeat, first twist off all the legs.

FIGURE 8.7b Pull the top shell off and reserve for stock.

FIGURE 8.7c Remove the feathery gills from the body.

FIGURE 8.7d Pull apart the body to expose the meat from the inner chambers.

FIGURE 8.7e Remove the meat with a small knife.

FIGURE 8.7f Crack the claws with the back of a knife or a nutcracker and remove the meat with a small knife.

SHELLFISH FABRICATION

Except for blue crabs, most species are sold fully cooked. Whole cooked crabs should be fabricated and used immediately. Commercially, crabmeat is removed with a variety of tools including small knives and compressed air.

OPENING CLAMS AND OYSTERS

To serve clams or oysters on the half shell, they first have to be opened manually. If they have been moved and jostled around, they are usually harder to open. Putting them on ice before opening will allow them to relax and loosen the muscle enough to at least get the knife between the shells. The shape of the clam or oyster determines how it should be opened. Clams have a hinge on the bottom that curves sharply around on one side and is straight on the other. The straight side will fit comfortably into your hand, allowing for the knife to be pressed into the other side.

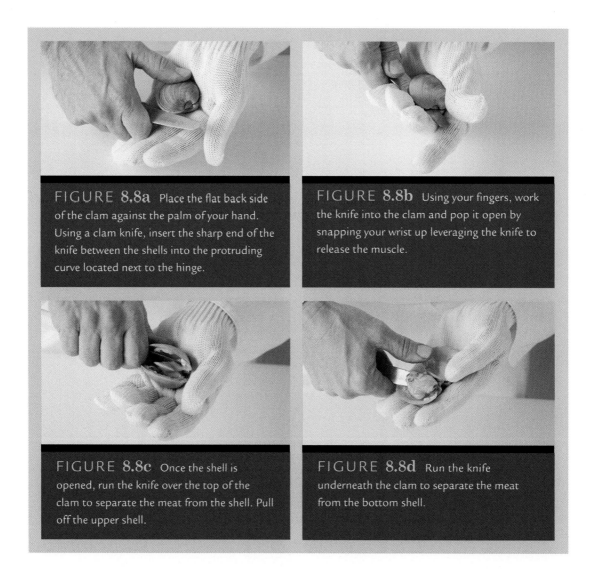

FIGURE 8.8a Place the flat back side of the clam against the palm of your hand. Using a clam knife, insert the sharp end of the knife between the shells into the protruding curve located next to the hinge.

FIGURE 8.8b Using your fingers, work the knife into the clam and pop it open by snapping your wrist up leveraging the knife to release the muscle.

FIGURE 8.8c Once the shell is opened, run the knife over the top of the clam to separate the meat from the shell. Pull off the upper shell.

FIGURE 8.8d Run the knife underneath the clam to separate the meat from the bottom shell.

FIGURE 8.8e Check the clam for any shell fragments.

Oysters come in varying shapes and can be difficult to open. The cup shape part of the shell, or what looks to be the stronger shell, should be held in the palm of your hand (see Fig 8.9a). This will leave the thinner and weaker half of the shell on top. Once the oyster is positioned properly, insert the tip of the knife into the hinge and leverage the top opened (see Fig 8.9a). Some oysters are very difficult to open, but realizing the importance of leverage, not strength, will make the process easier. For safety, wear a cut-resistant or chain link glove because dull shellfish cuts are slow to heal.

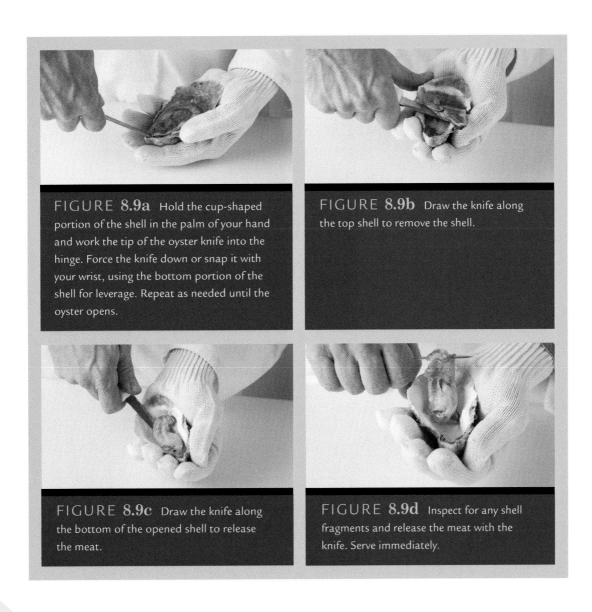

FIGURE 8.9a Hold the cup-shaped portion of the shell in the palm of your hand and work the tip of the oyster knife into the hinge. Force the knife down or snap it with your wrist, using the bottom portion of the shell for leverage. Repeat as needed until the oyster opens.

FIGURE 8.9b Draw the knife along the top shell to remove the shell.

FIGURE 8.9c Draw the knife along the bottom of the opened shell to release the meat.

FIGURE 8.9d Inspect for any shell fragments and release the meat with the knife. Serve immediately.

MUSSEL FABRICATION

Because most mussels are cultured and raised on ropes suspended from floats, they are very clean and often involve little if any cleaning other then pulling off the fine threads or beard located close to the hinge (see Fig 8.10).

FIGURE 8.10 Removing the beard from a mussel.

SEA URCHIN FABRICATION

Sea urchin roe can be obtained by removing it from the sea urchin or purchasing it from specialty purveyors. In Japanese it is known as *uni* and is carefully packaged on wooden trays.

FIGURE 8.11a The top portion of the sea urchin should be scraped with a fish scaler to remove the pins.

FIGURE 8.11d The black substance is the viscera and the orange is the edible roe.

FIGURE 8.11b, c Insert the scissors into the hole on top of the urchin's shell and cut the top off.

FIGURE 8.11e Remove the black viscera, being careful not to disrupt the delicate roe.

FIGURE 8.11f Remove the roe using a spoon; cook or serve immediately.

COOKED SKATE WING FABRICATION

Skate wings can be filleted raw or cooked. Place the knife flat on the bone/cartilage and remove the "fillet" which is very delicate but stringy. The tough and slimy skin is easier to remove once it is cooked. Repeat this procedure for both sides.

Skate is best when fabricated after it has been out of the water for a day or two. If it is too fresh, it will curl and be tough when cooked.

SHRIMP SHELLS

Shrimp lovers know that before you can enjoy this wonderful crustacean you must first remove the shell and vein. Shrimp shells are often discarded, which is wasteful because they can easily be transformed into complex sauces, stocks, and bisques that are delicious and profitable. Because the shells freeze well, they should be saved until there are enough for stock, which can be made by roasting the shells along with a mixture of carrots, celery, onions, and tomato paste. When caramelized, deglaze with brandy, cover with water, and simmer for 1 to 2 hours, then strain. In addition to sauces and soups, it can be reduced and used as a base.

There are many devices on the market that will speed up the shelling process. Hard plastic implements that are tapered and curved on top allow you to run the tip through the upper vein area, pushing to remove both the shell and the vein in one motion.

Most professionals forego the fancy devices for a simple paring knife which will easily slice through the back of the shrimp to access the vein.

PEELING AND DEVEINING

Minimizing the steps and motions to peeling and deveining shrimp quickly can be accomplished by developing a technique and sticking to it. If each shrimp is fabricated using the same motion, your body will develop a specific muscle memory for the task, greatly increasing your speed and efficiency.

BUTTERFLYING A SHRIMP

There are many reasons to butterfly a shrimp. Oftentimes it is done to remove the waste vein, which on wild shrimp can be large and sandy. Butterflying also makes the shrimp appear larger by giving it more surface area for breading and cuts down on the cooking time. For deep-fried shrimp, this decreased cooking time results in a perfectly cooked shrimp, whereas other techniques like baking or sautéing

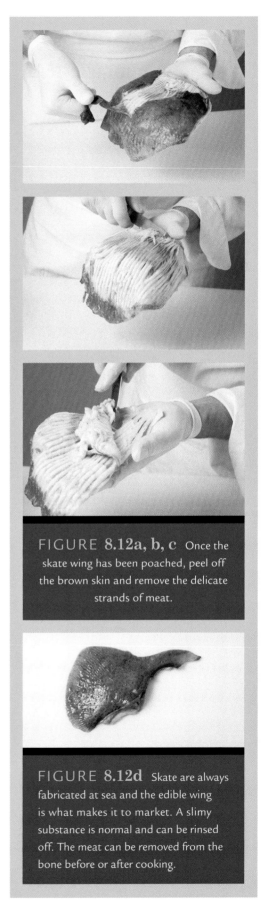

FIGURE **8.12a, b, c** Once the skate wing has been poached, peel off the brown skin and remove the delicate strands of meat.

FIGURE **8.12d** Skate are always fabricated at sea and the edible wing is what makes it to market. A slimy substance is normal and can be rinsed off. The meat can be removed from the bone before or after cooking.

SHELLFISH FABRICATION **183**

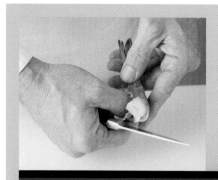  

FIGURE **8.13a, b, c** Place a small paring knife in your dominant hand, and pick up the shrimp in your opposite hand. Grasp one side of the shrimp's "feet" between your thumb and forefinger of the dominant hand and work the shell up and over the back in one complete motion.

FIGURE **8.13d** Once the shell has been removed, grasp the tail portion between your thumb and forefinger and pull it straight off.

FIGURE **8.14a, b, c** To remove the vein from the shrimp, make a 1/4-inch deep cut at the head of the shrimp just below the vein. Cut all the way down to the tail and pinch out the vein.

FIGURE **8.15a** To devein the shrimp without cutting it open, insert a skewer just above the tail section and underneath the vein.

FIGURE **8.15b, c, d** Pinch the vein and pull it straight down and out.

Butterflying a shrimp from the back

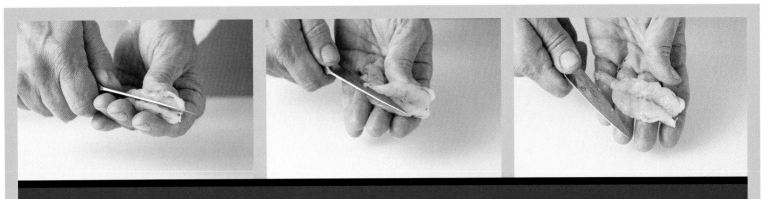

FIGURE 8.16a, b, c To butterfly the shrimp from the back, cut three quarters of the way through the shrimp from the head to the tail.

may cause the shrimp to overcook. Most shrimp are butterflied from the back giving it the traditional fan shape (see Fig 8.16a through 8.16c). The other technique is to open it up from the underside, resulting in a spoon shape which is perfect for stuffing (see Figs 8.17a through 8.17d).

Butterflying a shrimp from the bottom

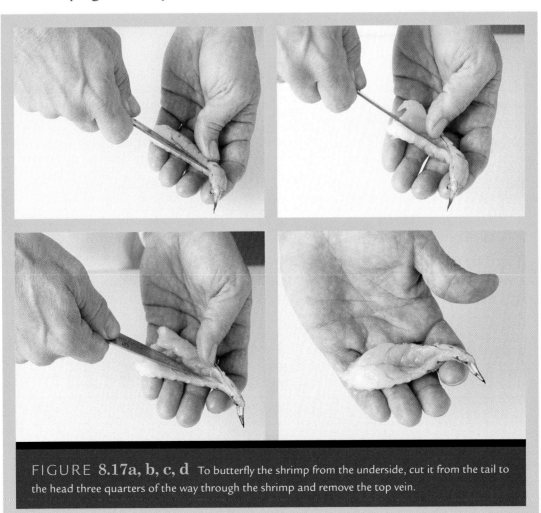

FIGURE 8.17a, b, c, d To butterfly the shrimp from the underside, cut it from the tail to the head three quarters of the way through the shrimp and remove the top vein.

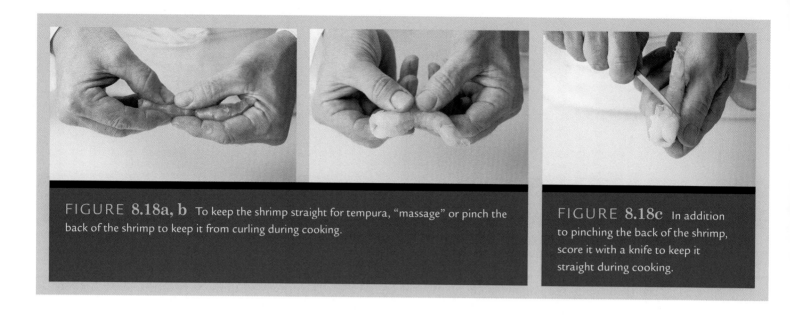

FIGURE 8.18a, b To keep the shrimp straight for tempura, "massage" or pinch the back of the shrimp to keep it from curling during cooking.

FIGURE 8.18c In addition to pinching the back of the shrimp, score it with a knife to keep it straight during cooking.

TOOLS OF THE TRADE

The objective of fish fabrication or butchery is to remove as much usable fish from the bones as possible with the least amount of effort. Fish butchers are rare breeds who work in cold, wet, slimy, strenuous conditions. A lot of the catch is filleted on the rolling seas by machine or hand, where skilled butchers and their knives cut the perfect sashimi, steak, or fillet. Carbon steel knives with wooden handles have been used on boats and in kitchens for centuries. Metals and handle materials have evolved allowing for durability while adhering to proper sanitation principles. All steel contains iron and carbon with smaller amounts of sulfur, phosphorus, manganese, and silicon. The

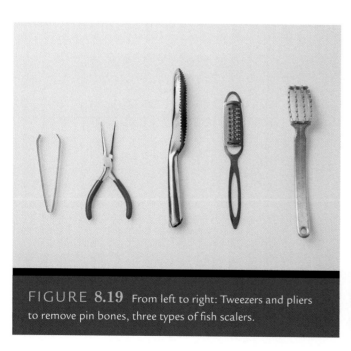

FIGURE 8.19 From left to right: Tweezers and pliers to remove pin bones, three types of fish scalers.

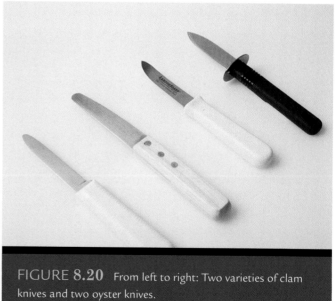

FIGURE 8.20 From left to right: Two varieties of clam knives and two oyster knives.

FIGURE 8.21 A collection of fish knives from left to right: flexible smoked salmon slicer, large fillet or steak knife, general utility or chef's knife, Japanese fish knife, flexible fillet knife, medium-size fillet or steak knife, ridged fillet knife.

more carbon in the steel, the harder it is. Traditional carbon steel knives discolor and tarnish when exposed to moisture or highly acidic foods but keep an edge for a longer period of time. Advances in stain-resistant materials such as chromium and molybdenum increase the strength, hardness, and rust-resistant qualities while still allowing for a superior edge. The trick to keeping knives sharp begins with the type of steel and ends at the stone or grinder.

Stones are made from synthetic materials, ceramic, natural materials, and some are even embedded with diamonds. Synthetic varieties have grits ranging in smoothness from 800 to 8,000. Lower numbers are coarser, and establish the edge, while higher numbers are finer, and remove scratches and create a mirror finish. Diamond stones are often not recommended for superior quality knives because they will remove too much material. Natural stones are very expensive, limited in supply and typically used by craftsmen and woodworkers. Water is the lubricant of choice because it is free, clean, and available, but oil can be used successfully. Wheel grinding machines and belt-sharpening devices give a quick edge but erode more material, cutting down on the life of the knife. Oftentimes choosing the right system for sharpening will be based on the quality of the knife and the time available to sharpen it. Blades containing higher amounts of stainless steel benefit from the machine method, whereas a superior Japanese sashimi knife should be sharpened by hand.

JAPANESE STYLE KNIVES

In the traditional style of Japanese knifemaking, all knives have an outside edge that is beveled. This means that unlike western style knives which are sharpened on both sides, only one side of a Japanese knife is sharpened. The other side of the blade is flat or concave to facilitate sharpening.

Modern Japanese knife producers use many of the same metallurgy and forging techniques handed down from the Samurai. Most come with a factory edge that still needs to be "polished" using a very fine water stone, not oil. All Japanese knives are designed for a specific purpose such as slicing fish or vegetables. Most all-purpose knives originate in the western kitchen.

HONYAKI KNIVES

- Made entirely from high-carbon steel
- Very time consuming to make and the most expensive
- Steel is very hard and brittle
- Highest quality, but difficult to sharpen

KASUMI KNIVES

- Forged from dual metals
- Steel is not very brittle and knives are easy to sharpen
- Moderately priced

KNIFE SHARPENING

PREPARING THE STONE

Water stones are the best choice for sharpening knives to be used in any food preparation. All composite stones should be submerged in water before use to allow them to absorb water. Once the stone is hydrated it needs to be kept slightly wet during the entire process to form the muddy residue needed to hone the edge. To keep the stone stable and wet during the sharpening, place a block of wood or a brick in a long hotel pan and cover it with a damp cloth. Place the stone on the damp cloth and use a small squirt bottle full of water to keep it wet. Three-sided stones, a luxury for many, are worth it if you do the volume or have deep pockets. Japanese stones are reasonable and work well with this system.

ESTABLISHING THE PROPER EDGE

The proper sharpening angle is determined by the density of what is being cut. For example, an ax that must go through hard wood is sharpened with a bevel of between 25 to 30 degrees. A knife that cuts through soft items such as vegetables or fish requires less of a bevel and is typically sharpened between 10 to 20 degrees. Establishing and maintaining the edge throughout the life of the knife will ensure many years of consistent service. Many knives come with what is referred to as a factory edge. This edge should still be sharpened using a stone with a grit appropriate to the steel used to make the knife.

To establish the proper angle of the edge, place the side of the knife flat on the stone and lift it up to 10 to 20 degrees depending on what is being cut and the specific type of knife. Keep this angle consistent throughout the sharpening and steeling for a consistent edge.

SHARPENING THE KNIFE

There are many ways to sharpen a knife. Some techniques involve moving the knife in small circular motions on each side to establish an edge, whereas others call for the knife to be sharpened on alternate sides one after the other.

The method illustrated here is used by most Japanese chefs who for centuries have had a reputation for keeping their knives razor sharp. It involves a repetitive back and forth motion from tip to heel for about 10 strokes on each side. Various sharpening stones, from coarse to smooth are used depending on the type of knife and quality of the edge, a process that takes time and patience.

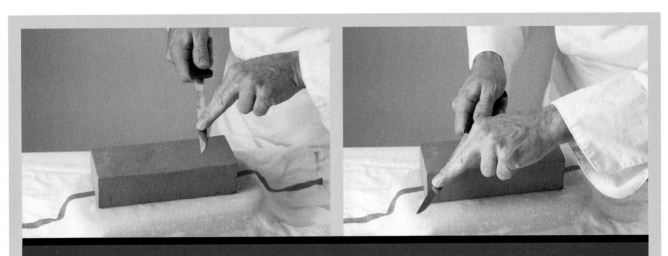

FIGURE **8.22a, b** Begin sharpening by placing the tip of the knife at one end of the stone and follow through with a consistent stroke and angle to the heel of the knife. Pull the knife back toward you from the heel to the tip. Continue this back-and-forth motion on one side for about 10 seconds.

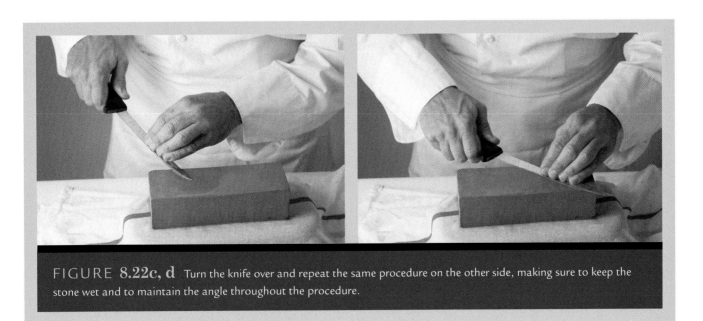

FIGURE **8.22c, d** Turn the knife over and repeat the same procedure on the other side, making sure to keep the stone wet and to maintain the angle throughout the procedure.

(*continues*)

(continued)

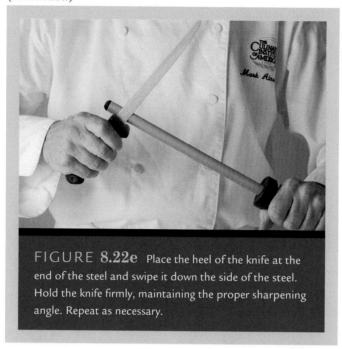

FIGURE 8.22e Place the heel of the knife at the end of the steel and swipe it down the side of the steel. Hold the knife firmly, maintaining the proper sharpening angle. Repeat as necessary.

FIGURE 8.23 The coarse stone in the background is used to establish or restore the edge of a dull knife. The two-sided stone in the foreground ranges from a coarse grit of 1,000 to a fine grit of 3,000 and is used to maintain an edge.

AQUACULTURE

Aquaculture has been practiced worldwide in one form or another for centuries. Lowland fields and rice paddies would often be flooded by strong rains and monsoons. With these floods came an abundance of fish and aquatic organisms which flourished in this vegetation-rich environment. Once the rainy season ended and the water receded, the fish and their offspring would be harvested. Eventually it was realized that by leaving the smallest of the fish in these ponds, they would grow, spawn, and the cycle could be continued and managed as long as there was water and food available for the fish. In areas close to the sea, with concentrations of tidal ponds, and lagoons that were affected by the ebb and flow of tides, traps would be installed that would block the exit of these fish.

The next steps in aquaculture were to gather up small fish (fry) from rivers, streams, and oceans and transport them to controlled ponds of fresh or brackish water where they were allowed to grow before being harvested. This technique was widely used in China with the common carp. It wasn't until the eighteenth century that eggs and sperm from fish ready to spawn were removed and spawned in controlled conditions before being moved to tanks or ponds for cultivation. Initially these were all freshwater fish, but in the twentieth century, new techniques were developed to breed saltwater species.

With these new techniques has come expansion of farming high value seafood such as sea bass, salmon, shrimp, and grouper, a practice that has dropped prices, but caused controversy for its possible environmental impact.

FISH FARMING TECHNIQUES AND METHODS

AQUAPONICS

Aquaponics is the process of combining aquaculture and hydroponics. This process utilizes the waste products and water from the fish for plant nutrients. The plants act as filters for the water, which is purified and recycled back into the tanks. No chemical fertilizers, pesticides, or medications are needed for the plants or fish. The water needed in this process is about 95 percent less than normal fish farming and 1/10 the water required for normal vegetable growing. Aquaponics is viable for both large- and small-scale operations and can be accomplished with both natural and artificial light sources. Many species of fish, herbs, and vegetables can be grown using this technique, which is an ideal sustainable form of both aquaculture and agriculture for the future.

SEA RANCHING

Sea ranching is an aquaculture technique that is becoming controversial in many parts of the world. Large schools of high-value fish such as tuna are captured through a well-orchestrated multi-boat process. The fish are rounded up in what amounts to an underwater cattle drive as they are herded out to sea and into waiting circular nets that float on the surface. These large netted enclosures are carefully towed out to sea where the fish are fed and fattened up over a period of months. The controversy over this technique arises because large amounts of fish are rounded up, often without regard to quotas, and in the process of the trapping there are casualties. Another problem is that many fish are carnivorous and require a diet of smaller fish such as herring and sardines. Because carnivorous fish do not metabolize carbohydrates well, the amount of food they consume exceeds the amount they produce at a rate of up to five to one. This oftentimes illogical number is consistent with most meat-eating farmed fish, putting a strain on many of the smaller species used to create fish protein and meal.

CAGES

Tuna as well as salmon are held or raised in net-like enclosures anchored to the sea bottom. Large amounts of fish can be monitored remotely with cameras and automatic feeders. Because of the large concentration of fish and feed, pollution and algae bloom is possible. Additionally, there is fear that interbreeding and/or the spread of disease will eventually disrupt the wild species.

RECIRCULATION SYSTEMS

This method is a lot like a large, round aquarium. Located both indoors and outdoors, treated water is constantly recirculated through the system to raise a variety of species including salmon, bass, and even sturgeon.

RACEWAYS

These systems resemble long, deep lanes or channels through which diverted stream or river water is pumped. The fish, typically trout, are able to swim in these channels;

their wastewater is treated before being returned to the source. Regulations are in place in many states minimizing environmental impact.

PONDS

Trout, catfish, shrimp, and tilapia are examples of species raised in ponds. They are built in a variety of shapes and materials, including concrete and earth, and have the advantage of being placed near water sources both inland and in coastal areas. A constant supply of water is important to the delicate balance of this crowded environment and surrounding ecosystem.

ALGACULTURE

Algaculture refers to the controlled farming of algae, or seaweed. Common to the Asian diet, it is highly nutritious, containing vitamins A, B1, B2, B6, C, and niacin. Additionally, the omega-3 fatty acid docosahexaenoic acid (DHA) can be manufactured from algae and is used as a nutritional supplement. Many varieties of algae are used as natural dyes; livestock feed, fertilizer, or to control pollution in wastewater plants. It can also be made into bio-diesel. As its health benefits and versatility continue to be studied, algaculture will certainly increase in the coming years.

SHRIMP FARMING

For over thirty years, shrimp farming has been on the increase worldwide. Limited supply of wild shrimp, high profits, and relatively cheap start-up costs in depressed economic regions have fueled this expansion. Much of the growth is in developing countries such as Thailand, Indonesia, India, the Philippines, and Ecuador, and has brought jobs and economic gains, as well as environmental degradation. The hope of quick profits has led to overexpansion and unrealistic goals, without proper capital and controls being invested in environmental safety. In question is whether this young industry can sustain itself amidst increasing world pressures to mitigate and end disruption of wet lands and control possible disease outbreak to wild shrimp.

The shrimp that we enjoy fried, sautéed, and grilled are produced in a wide variety of methods. Intertidal ponds that allowed shrimp to enter naturally was one of the first methods used in shrimp farming. Because shrimp trapped in these environments were not prone to fertilization, and their natural food supply was not present, these ponds proved less than efficient. This led to closed pond systems that are seeded with post-larvae shrimp seed and are fed and fertilized to encourage accelerated growth. Most farms today use wild shrimp for production of seed that comes from nursery ponds or hatcheries. Farmed shrimp can be matured and spawned in hatcheries, but production, volume, and quality is lower than from wild-caught spawning shrimp, which carry a very high price tag. It is assumed that the high price of wild-caught spawning shrimp will intensify research to develop quality farm-raised brood stock, eliminating the need for wild shrimp. Most farms purchase larvae from hatcheries that are either in their first stage of development or in post-larvae stage which are considered juveniles. Black tiger shrimp *(Penaeus monodon)* account for the majority of farm-raised species worldwide. They are tolerant to low salinity and can typically reach harvesting size in about 3 to 5 months. Depending on the climate and temperature, approximately one to three crops of farmed shrimp can be grown per year.

Most shrimp farming takes place in intertidal zones and wetlands that support a wide range of important aquatic plants and animals. For many years these wetlands were viewed as undesirable for farming or urban development because of the high cost of reclamation. With the boom in the shrimp market and the technological advancements associated with it, local governments encouraged private development of these areas as a way to stimulate economic growth. At the time, scientists were just realizing the ecological value of these areas that include mangrove forests, salt flats, brackish creeks, rivers, lagoons, and a variety of sea plants, grasses, and marine life. Now recognized as having a very important role in the overall health of coastal areas, these areas act as a barrier to protect the shoreline from storms and floods and reduce erosion. They are critical to the filtering of organic nutrients and coastal runoff, act as nurseries for a variety of marine life, and are a part of a very important balance which must remain intact.

Governments are now realizing the importance of these wetlands to the overall health of coastal areas, and are beginning to regulate their use. Mangrove forests can be beneficial to shrimp farms located nearby because they can successfully remove, filter, and clean discharges and effluents produced by the shrimp farms. This can be a delicate balance; the quantity of the discharge has to be efficiently gauged and monitored, because large amounts of concentrated organic material, chemicals, and antibiotics can easily disrupt the ecosystem of the wetlands.

Because of the large amounts of seawater needed to raise the shrimp, farms are situated near the ocean. To combat evaporation, new water is constantly replenished through tidal flow or by the introduction of fresh water to a salinity rate of 15 to 20 ppm. This usage of fresh water has been problematic in areas with an insufficient groundwater supply for agriculture and domestic usage. After harvesting, a large amount of chemically sensitive discharge is released into the environment. This needs to be handled effectively with the local ecosystem in mind.

Adaptable to varying salinity rates, black tiger shrimp are ideal for freshwater farming. In their first stage of growth, shrimp are raised in salt water that is trucked to the site. Slowly, the salinity is reduced to nearly zero for the majority of the growth cycle. This method of production is not without its problems. Management of the salt and effluent is critical to avoid polluting freshwater streams. Ponds should be constructed on clay soil to avoid salt and chemicals from leaching into the ground water, and proper feed management, which accounts for at least half of the operational cost, should be monitored and studied to mitigate algae bloom.

Disease associated with shrimp farming has not been found to be harmful to humans, but has plagued the industry and undermined profits. The worst outbreaks have been viral in nature with at least fifteen different viruses identified. Antibiotics and proper sanitation practices have proven effective against the viruses and assorted bacteria. There is growing concern that the indiscriminate use of medication and chemicals will impact negatively on the shrimp and the environment. Eliminating diseases without the use of chemicals is problematic in view of the crowded and dirty conditions the shrimp inhabit. Without the use of modern pharmaceutical and sanitation practices, the delicate balance of a captive life cycle is impossible to sustain.

In captivity, shrimp eat food comprised of fish meal and fish oil made from species of fish not typically consumed by the majority of humans. Concern is growing that aquaculture's dependence on wild fish will cause depletion of stocks. Approximately three pounds of wild fish are needed to feed and grow one pound of shrimp. Managing and controlling this type of fishing will be important as the need and demand for fish meal increases. Future development of single cell foods, such as yeast, will also contribute to a sustainable industry, but more research into this science will need to be done.

It should be pointed out that the delicate flavor of shrimp is dependent on its diet and environment. From a culinary standpoint, there are many differences between farm-raised and wild-caught shrimp. The flavor of wild shrimp is far superior to the farmed varieties. Their intestinal veins are often full and must be thoroughly cleaned before use. Their shells have a different appearance and color, and the cooking times and textures are slightly different between varieties. Farm-raised shrimp have become such a mainstay on the table that many people have never even tasted a wild shrimp, or are aware of what they are purchasing or eating.

As the shrimp industry works hard to keep up with world demand, consumption of farm-raised shrimp will increase. Wild stocks of shrimp and fish used for meal will need to be continually monitored worldwide to avoid overfishing, and research should continue in the areas of feed development and diseases. With time the industry should be able to balance itself amongst its many obstacles giving chefs and consumers, a safe, eco-friendly, delicious product.

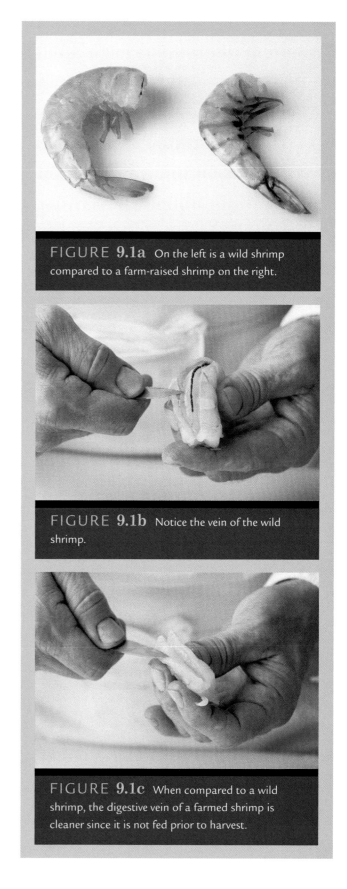

FIGURE **9.1a** On the left is a wild shrimp compared to a farm-raised shrimp on the right.

FIGURE **9.1b** Notice the vein of the wild shrimp.

FIGURE **9.1c** When compared to a wild shrimp, the digestive vein of a farmed shrimp is cleaner since it is not fed prior to harvest.

SANITATION: SAFETY AND SANITATION, STORAGE AND HANDLING

Ice is a complicated interaction of water and temperature that slows down and inhibits bacterial growth. For centuries it was cut out of frozen lakes and rivers and stored nearby under burlap and straw to be used long after winter's end. Writings during Roman times mention emperors having ice carted down from the Alps so they could enjoy a bowl of ice cream on a hot summer night. Prior to refrigeration, people ate what was local and available and much of the meat and fish was preserved using methods handed down through generations. Freezing technology, motorized sea vessels, and the railroad were several important outcomes of the Industrial Revolution that would change the seafood industry forever.

With the advent of new technologies, large amounts of fresh seafood could be shipped via rail to the large metropolitan centers of the world. The once small fishing villages along the coasts of industrialized nations immediately experienced an increased customer base and the rush to meet the demand was on.

This race to industrialize the industry continued with Clarence Birdseye's invention in the mid-1920s of the quick-freeze machine which rapidly froze packaged fish under pressurization. Birdseye continued to improve on his invention and eventually sold his company, General Seafood, to Goldman Sachs, which later changed the name to General Foods Corporation.

The merging of all these technologies occurred after World War II with the development of the large factory ships. These mammoth vessels could stay out for months at a time, catching, filleting, and freezing their haul until their holding capacity was exhausted. This large-scale fishing began to take its toll on the seas, prompting the establishment in 1949 of the International Commission for the North West Atlantic Fisheries to control these massive and destructive fishing practices. To this day, advancement in sonar and satellite tracking continue to guide the large ships to their target where they scoop up and deplete the ocean of its remaining supply.

Fish is high in water and protein and must be properly chilled immediately after landing and gutting. The body temperature of certain species such as tuna can rise to 86°F/30°C during the struggle of the catch. These high temperatures are best brought down rapidly with a slurry of flaked ice and seawater. After that, the fish should be packed in ice until the boat reaches shore. The slurry might seem extreme, but for some fish the high market price warrants the extra attention to detail. Once the fish reach shore they are packed in waxed boxes, iced, and shipped by truck, plane, or train. Temperature fluctuations at this point are detrimental to the fresh cargo, and speed to market is critical. These shipments typically head to large markets or distributors before being consigned to grocery stores and fishmongers. Once at their final destination, they are kept in walk-in refrigerators at 40°F/4°C or below until removed to the display case for sale.

HOW FRESH IS FRESH FISH?

When we gaze into the display case of our local supermarkets in Boston or Seattle, we see beautiful red snapper or grouper packed in crushed ice ready for the grill or oven. The fish's migration from ocean to vendor undoubtedly has entailed its capture in the first hour of a fishing trip that lasted 48 hours far out in the Gulf of Mexico. Once hauled out of the ocean, it is gutted and chilled, using all modern methods available to the crew. Upon returning to the dock, the fish is placed in ice and transported by truck for the 30-hour trip to a Boston wholesaler. Arriving late on a Saturday night, the entire off-loaded product is transferred to refrigeration where it sits until the start of business on Monday morning. By Monday afternoon it is swimming in ice ready for purchase, which, with any luck, will happen in the next day or two. From the time it is caught to the time it is consumed, over a week has passed. This is common practice for the industry, but consumers often mistakenly assume that their fresh fish is just hours or days old.

Typically, fin fish and saltwater shrimp (with their heads removed), properly handled and chilled, will have a shelf life of between 10 and 12 days. Freshwater shrimp, crab, and crawfish have a much shorter shelf life of about 3 to 4 days.

TIME AND TEMPERATURE

The consequences of time and temperature fluctuations throughout the distribution chain have been found to be cumulative and can be crucial to product quality. Temperatures should be consistently monitored and logged using available computer software and Hazard Analysis Critical Control Point (HACCP) guidelines. It is well known that bacteria responsible for quality degradation of fresh fish grow very slowly at temperatures between 32° and 41°F/0° and 5°C; shelf life can be extended by maintaining a cold and consistent environment. A fresh fish with a known shelf life of 10 days when stored in ice at 32°F/0°C decreases in shelf life to under five days when held at 41°F/5°C. Individual species of fish and shellfish, fat content, and other factors all contribute to specific rates of spoilage.

Ice has long been used to keep seafood fresh. It has a very extensive cooling capacity and has the ability to avert surface dehydration of the fish as it melts; the water is able to transfer the fish's body heat to the surface of the ice more efficiently than air. Ice and water slurries are very efficient and practical ways to chill down fish, especially large species such as tuna and swordfish.

PROPER SHIPPING CONTAINERS

Another important tool to keep seafood at consistent temperatures is the use of insulated boxes and containers during holding and transport. Although it is the ice's function to keep the fish chilled, it is the insulated container that keeps the ice from melting. These containers can be cost effective by increasing the amount of product that can be shipped, by decreasing the amount of ice weight (and therefore the amount of water needed), thus lowering shipping cost.

SUPER-CHILLING

Super-chilling is the method of maintaining the temperature of seafood below 32°F/0°C but above freezing. Although this process is able to drastically increase the shelf life, it is difficult to accomplish in large volume because different species freeze at slightly different temperatures depending on muscle structure and moisture content of the fillets.

ICE

There are many varieties of ice commercially available for cooling seafood. They are produced in an assortment of shapes and sizes and can be made from fresh and salt water. Saltwater ice has a lower melting point (depending on the salt content) than freshwater ice, but is not as stable because the salt can separate from the ice. Salt may also have a detrimental effect on the seafood if the fillets absorb it. Ice varieties are differentiated by their specific surface area and physical characteristics.

BLOCKS
- Sold in a variety of sizes; it is ground up or chipped before being used to chill seafood
- Take up less storage area during transport and melt slower than other varieties

FIGURE **10.1a** Cubed ice versus crushed ice.

FIGURE **10.1b** Cubed ice is not the best choice when icing down seafood; its edges can actually bruise the delicate flesh of the fish.

FIGURE **10.1c** Crushed ice is good for seafood storage because it enrobes the product, maintaining a consistent temperature.

CRUSHED ICE (SEE FIG 10.1a THROUGH 10.1c)

- Compensates well for heat loss because of its volume
- Sharp edges can damage delicate product
- Lasts longer than flaked ice

FLAKE ICE

- Enrobes the fish very efficiently in containers, allowing for uniform distribution of temperatures
- Does not damage the delicate product
- Takes up more space in the containers, allowing for less product

COMMERCIAL FISH IS FROZEN IN A VARIETY OF WAYS

Often referred to as IQF, or individually quick frozen, blast freezing is a technique that circulates −25° to −40°F/−32° to −40°C air over fish fillets or seafood moving through a tunnel-like chamber on racks or conveyor belts. They are "sealed" in a protective coating of ice and remain separated, allowing for better portion control. Another benefit to IQF products is that they thaw quickly and have limited water retention.

CRYOGENIC FREEZING

Building on the techniques learned over a century of freezing technology, cryogenics involves items subjected to temperatures below −238°F/−150°C. In the seafood industry, individual products are placed in a spray or dip of carbon dioxide or liquid nitrogen at a temperature below −238°F/−150°C. This accelerated freezing develops a smaller ice crystal, resulting in minimal flavor and water loss while maintaining the structure and firmness of the meat. Fish frozen by this method is easier to prepare with high-heat cooking methods such as sautéing and grilling. The limited water loss allows for better caramelization and superior flavor and texture.

Driving the industry's quest for the perfect piece of frozen seafood is the consumer's renewed interest in taste at a reasonable price. Gone are the days of the poorly frozen blocks of fish and the stale-tasting fish sticks. Current advances in marketing and technology have been affected by the realization that the consumer understands quality and is willing to pay the price for a premium product.

FROZEN SUSHI

In the past few years, sushi has become popular in the United States. Grocery and convenience stores now sell trays of pre-made California rolls alongside hotdogs and roasted chicken. Chefs and consumers alike are confused as to the quality and correct preparation method of this apparently simple food.

Sushi describes many types of food, not only raw seafood. However, for those sashimi varieties that call for raw product, it is important to understand Food and Drug Administration (FDA) regulations requiring that all fish to be consumed raw must first be frozen to kill potential parasites. Tuna, considered a very clean fish free from worms and parasites, is a possible exception.

It is estimated that over half of the sushi in the United States has been previously frozen. This surprisingly high number is due to health concerns as well as simple market conditions of supply and demand. Tuna and other species of fish are seasonal and, because of increased consumption, are not always available fresh. Many consumers would be surprised to learn that the thin slice of ten-dollar tuna draped over a finger of rice had recently been taken out of the freezer. Japanese consumers, on the other hand, understand the fresh and frozen market and realize there is little difference between the two products as long as the freezing methods are correct.

Technology has advanced mightily since Charles Birdseye froze his first package of green beans. Today fish is frozen at −238°F/−150°C using patented techniques able to preserve flavors by eliminating the moisture between the cells, allowing the product's structure and texture to remain intact. Thawing of these super-frozen items is accomplished slowly, often in specially designed thawing containers, so they may come up to temperature in a controlled environment.

HOW TO FREEZE YOUR OWN FISH

Freezing your own seafood is relatively easy as long as you follow a few simple procedures. All items to be frozen should be rinsed clean and dried. This will reduce possible surface contamination and eliminate excess moisture and consequent freezer burn. Before freezing, fish should be cut into fillets, steaks, or other suitable smaller portion sizes of no more than 1 to 2 inches thick. Large loins take much longer to freeze than smaller pieces; the more time it takes, the more the cell structure will be affected. Thoroughly seal the fish in a plastic bag, making sure to remove as much air as possible. Vacuum-pack bags, which work well to remove all the air, should not be used to preserve or extend the life of fresh foods because anaerobic bacteria can still be present in these products. Proper sanitation and refrigeration procedures are very important when vacuum packaging. Once the air is out of the bag, it should be rewrapped in foil or plastic wrap and placed in a pan or other sturdy container several inches high, filled with cold water at least an inch or two above the fish, and placed in the freezer. This frozen barrier will ensure that no oxidation can occur and it will allow for a slow thawing.

STORAGE

Most home freezers are full of unidentifiable food of indeterminate age. When storing any product with a specific shelf life, it is important to label and date it. After which it should be stored at a minimum of 0°F/–18°C or colder. Depending on the species, it is recommended to freeze fish in a conventional freezer for no more than 3 to 10 months. This wide range is generally based on fat content of the product. Oily fish such as salmon and sable fish should be on the low end, whereas lean species like flounder and tilapia can last almost a year.

THAWING FROZEN SEAFOOD

Seafood is most efficiently thawed under refrigeration. The time it takes will depend on the thickness and temperature, but should be accomplished in 12 to 24 hours. Alternatively, frozen product can be thawed under cold running water. Because seafood will absorb moisture, avoid running water directly on the fish; instead, thaw it in its packaging.

With commercial deliveries easily available daily for most restaurant and grocery stores, prolonged deep freezes should occur only infrequently. The quality of a product dug out of the bottom of the freezer will never be the same as that of the iced wild-caught day boat cod that you picked up from the local fishmonger.

HOLDING TIMES FOR FRESH AND FROZEN SEAFOOD

PRODUCT	UNDER REFRIGERATION 40°F	FREEZER 0°F
Fresh seafood	1 to 3 days depending on source	3 to 10 months depending on oil content
Cooked seafood	1 to 3 days depending on freshness and how fast it was chilled	1 to 2 months
Frozen seafood	Only refrigerate to thaw	2 to 4 months
Smoked fish	1 to 2 days	Purchase only what you need; loses quality in the freezer
Dried and cured fish	6 months depending on species and how it was cured	Not recommended
Seafood soups and stew	1 to 3 days	3 to 6 months

STORAGE GUIDELINES

FIFO is an industry term that refers to the concept of first in, first out. Use up the old product before the new one. In theory, this simple practice is straightforward, but unless it is supported by expiration dates and proper labeling, it will not be successful.

REFRIGERATION

A refrigerator is only as good as its thermometer. Modern commercial and many home models have digital read-out thermometers built in that are extremely accurate and reliable. However, manual thermometers should also be used as back-ups to ensure a proper temperature between 34° and 40°F/1° and 4°C.

PURCHASING

Finding a reputable dealer for seafood requires knowledge of the product and industry. Long ago, your local fish store sold only species that were sourced locally, thus the supply chain was much shorter. This slowly changed with the advent of refrigeration and improved transportation. Today, seafood is a global market with an array of product available from Australia, Thailand, and Alaska. Eating what is local is still a good practice, because it is always fresher and supports the local economy. Regardless of your final choice, there are several things to look for when purchasing fish.

KNOWLEDGE

Ultimately, the person selling the product should be knowledgeable about its origin, how it is caught, and what its individual characteristics are. He/she should understand its flavor and preferred cooking method and nutritional properties, as well as whether it is wild-caught or farm-raised. Because most of the time purveyors will always tell you that everything is fresh, it's wise to question them about when their deliveries arrive and inquire about what has just come in. Developing a rapport with your fishmonger will pay huge dividends. Even grocery stores realize the importance of a satisfied customer and often have literature available to help the inexperienced attendant.

FOOD SAFETY AND HACCP

Proper storage of fresh and frozen seafood is critical to ensure its quality. Developing a relationship with your suppliers and asserting your knowledge about excellence will guarantee they're providing a quality product. Current industry guidelines dictate that all those in the supply chain follow a HACCP (hazard analysis critical control point) plan, which is a systematic preventative approach to identifying possible food safety problems. HACCP was begun in 1959 when Pillsbury Corporation collaborated with NASA in a program to systematically monitor food served in space. HACCP is now recognized worldwide as a leader of food safety in all venues. The program establishes standards for plant design, production, packaging, storage, and delivery to ensure food safety and prevent physical, chemical, and microbiological hazards. Individual HACCP plans are examined by the United States Department of Agriculture (USDA) or the local health department and are monitored by the processors. These inspections ensure

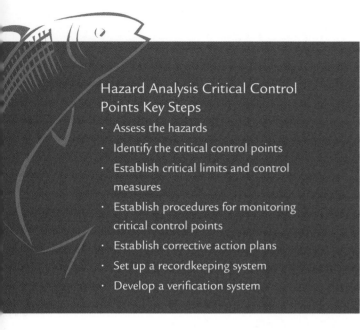

Hazard Analysis Critical Control Points Key Steps
- Assess the hazards
- Identify the critical control points
- Establish critical limits and control measures
- Establish procedures for monitoring critical control points
- Establish corrective action plans
- Set up a recordkeeping system
- Develop a verification system

that each of the "critical control points" or those areas most susceptible to food-borne pathogens meet the established guidelines and are tracked, evaluated, and recorded to guarantee compliance. Although the FDA has regulated the seafood industry for many years, it only recently has mandated the use of HACCP programs. Companies in the United States or abroad that handle, store, fabricate, shuck, eviscerate, preserve, pack, label, unload, hold, or prepare seafood destined for the United States must have an HACCP plan in place http://www.cfsan.fda.gov. The FDA has estimated that these regulations setting the standards in safe seafood-handling practices worldwide prevent up to 60,000 fish poisonings each year.

Unlike land animals, seafood is delicate. It does not have the muscle structure or fat and skin layer of pigs and cows. Its high protein and water content make it much more susceptible to pathogens than meat and should be handled within the HACCP guidelines. Compliance with these guidelines is an expense that gets passed on to the customer, but the reduced incidence of food poisoning is worth the price.

GOVERNMENTAL SEAFOOD REGULATIONS

The FDA is the agency that regulates public protection of the country's seafood supply. Under the USDA, seafood processors undergo yearly inspections to ensure compliance with product safety, wholesomeness, identity, and economic integrity. In addition, the FDA performs inspections of domestic seafood harvesting.

Notified of all seafood shipments into the country, the FDA has the authority to detain and hold product it feels is misbranded or adulterated. Its laboratories around the country scientifically test for a wide range of defects including microbial pathogens, chemical contamination, food and color additives, toxins, and pesticides. It also has authority to set tolerance levels for both natural and man-made contaminants in food and can take legal action against those who ignore these regulations.

Other regulatory programs that deal with seafood are the Salmon Control Plan and the National Shellfish Sanitation Program. The Salmon Control Plan is voluntary and provides control over plant sanitation and processing. The National Shellfish Sanitation Program was created to evaluate and control the sanitary production and harvest of fresh and frozen mollusks. Shellfish tags, a prime example of this program, enable the product to be traced back to the harvesting location if specific problems arise.

Unlike meat, which is carefully inspected and stamped by the federal government, seafood is largely left up to the processors and operators to oversee quality and follow correct HACCP guidelines. In addition to these governmental regulatory agencies, individual companies realize the importance of quality control. The foremost leader in the quest to serve fresh and safe seafood is Boston's Legal Seafood, which has set up a multi-tiered system to ensure the quality of its raw product. Inspections begin on the fishing boat, continue to a state-of-the-art processing and inspection facility, and end at the company's restaurants. This attention to detail is a prime example of industry taking charge of its own safety programs to the mutual advantage of company and customer.

FOOD-BORNE DISEASE

Food poisoning is a gastrointestinal disorder caused by consuming foods or beverages (including ice) that have been improperly cooked, handled, or stored. Sources of contamination that affect the food supply can be chemical, physical, or biological. Cleaning products are an example of chemical contaminants that may find their way into food. Physical contaminants include glass, hair, and dirt. Careless food handling such as an earring or foreign object falling into the food can cause injury. Biological contaminants account for the majority of food-borne illness. These include naturally occurring toxins found in plants such as wild mushrooms and rhubarb leaves. The predominant biological agents, however, are disease-causing microorganisms known as pathogens which are responsible for up to 95 percent of all food-borne illnesses. Microorganisms of many kinds are present everywhere, and most are helpful or harmless; only about 1 percent of microorganisms are actually pathogenic.

Food-borne illness caused by biological contaminants fall into two subcategories: intoxication and infection. Intoxication occurs when a person consumes food containing toxins from bacteria, molds, or certain plants and animals. Once in the body, these toxins are poisonous. Botulism is an example of intoxication. In the case of an infection, the food eaten contains large numbers of living pathogens that multiply in the body, attacking the gastrointestinal lining. Salmonellosis is an example of an infection. Some food-borne illnesses have characteristics of both an intoxicant and an infection. *E.coli* 0157:H7 is an agent that causes such an illness.

FOOD PATHOGENS

Pathogens responsible for food-borne illness are fungi, viruses, parasites, and bacteria.

Fungi include molds and yeasts which are more adaptable to various conditions than other microorganisms.

- Have high tolerance for acidic conditions
- Are more often responsible for food spoilage than for food-borne illness
- Are used in the production of cheese, bread, wine, and beer

Viruses do not actually multiply in food, but if, through poor sanitation practices, a virus contaminates it, consumption of that food may result in illness

- Hepatitis A is a virus caused by eating shellfish harvested from polluted waters
- Poor hand-washing practices spread viruses

Once in the body, viruses invade a host cell and reprogram it to produce exact copies; the copies leave the dead host cells behind and invade more cells.

Parasites are pathogens that feed on and play host to another organism. This host does not benefit from the parasite.

- *Trichinella spiralis* worms are known to infect hogs
- Anisakis roundworms are found in wild-caught fish including salmon
- Following proper cooking temperatures is an effective way of eliminating parasites in fish and meats

An example is the parasitic worm that exists in the larva stage in muscle meats. Once consumed, the life and reproductive cycle continue. When larvae reach adult stage, the fertilized female releases more eggs. These eggs hatch and travel to the muscle tissue of the host, and the cycle continues.

Bacteria are responsible for a significant percentage of biologically caused food-borne illness. To protect food during storage, preparation, and service, it is important to understand the classifications and patterns of bacterial growth.

Bacteria are classified by their requirement for oxygen:

- Aerobic bacteria require the presence of oxygen to grow
- Anaerobic bacteria do not require oxygen and can die when exposed to it
- Facultative bacteria are able to function with or without oxygen

Reproduction of bacteria takes place by means of fission. One bacterium grows and then splits into two bacteria of equal size. These bacteria divide to form four, the four divide to form eight, and so on. Under ideal conditions, bacteria reproduce very quickly and in great numbers.

- In about 12 hours, one bacterium can multiply into 68 billion bacteria, enough to cause illness
- Certain bacteria can form endospores that protect them from heat and dehydration
- Endospores allow bacteria to resume their life cycle if favorable conditions recur

Bacteria require three basic conditions for growth and reproduction:

- A protein source: the greater the amount of protein in the food, the greater its potential for food-borne illness.
- Moisture: measured on the water activity scale (Aw) measured from 0 to 1 with 1 representing the Aw of water. Foods with water activity above 0.85 support bacterial growth.
- Moderate pH: a food's relative acidity or alkalinity is measured on a scale known as pH that runs from 0 to 14. A moderate pH, which supports bacterial growth, is 4.6 to 10. Most foods fall within this range. Acidic foods such as vinegar and lemon juice have low pH and will inhibit bacterial growth.

Potentially hazardous foods are those that meet the three conditions necessary for bacterial growth. Examples of potentially hazardous foods are:

- Seafood
- Meats
- Poultry
- Tofu

- Cooked rice, beans, pasta, potatoes, sliced melon, sprouts, garlic, and oil mixtures
- Soft to mild cheese; certain dry cheeses do not contain enough moisture to support growth

Foods that contain pathogens in sufficient numbers to cause illness may still look and smell normal. Disease-causing microorganisms are too small to be seen with the naked eye, hampering detection. Because the microorganisms, particularly the bacteria that cause food to spoil, are different from the ones that cause food-borne illness, food may be contaminated and still have no off smell or odor.

Although cooking food will destroy many of the microorganisms present, careless food handling after cooking can reintroduce pathogens that are able to grow at an accelerated rate from the microbes that cause spoilage.

CROSS-CONTAMINATION

Many food-borne illnesses are the result of unsanitary handling procedures. Cross-contamination occurs when disease-causing elements or harmful substances are transferred from one contaminated surface to another.

STEPS TO REDUCE AND ELIMINATE CROSS-CONTAMINATION

- Practice personal hygiene, including hand washing and avoiding contact with food when diagnosed with a contagious illness or an infected cut
- Utilize separate cutting boards and work stations for raw and cooked foods
- Use proper sanitation solution and procedures when working with all food
- Store food correctly, placing cooked foods above raw foods or in separate locations
- Wear food-handling gloves for ready-to-eat foods

HAND WASHING IS ONE OF THE EASIEST AND MOST EFFECTIVE WAYS TO REDUCE FOOD-BORNE ILLNESS:

- Wash hands and forearms in hot soapy water for twenty seconds
- Water temperature should be hot, at least 110°F/43°C
- Wash hands after using the toilet and before handling food
- Wash hands after touching any surface or object that could possibly be contaminated

FOOD-HANDLING GLOVES

If you have ever observed the person who makes the pizza, pours you a soda, then makes change, all with the same pair of gloves on, you must wonder what purpose the gloves serve. Many states have glove laws that stipulate the wearing of gloves for all ready-to-eat foods. This means that if a foodservice worker is mixing a Caesar salad with their hands, they must wear gloves because the salad will not be cooked. On the other hand, if a foodservice worker is traying up raw chicken breast to be grilled, no gloves are needed.

Hand-to-food contact is one of the largest contributors to food-borne illness. This includes fecal and oral contamination, and transmitting bacteria through open wounds or handling food while sick. The purpose of the glove is to form a barrier between

Guidelines for Single-use Gloves for Food Handling:

WEAR GLOVES:

- When exposing hands to ready-to-eat food that will not be cooked
- To cover artificial nails or nail polish
- To cover an orthopedic support that is worn on your hand or wrist
- To cover a bandaged cut or sore

CHANGE GLOVES:

- Between specific preparations, washing hands between applications
- After washing hands, whenever beginning food preparation
- After touching or covering your mouth
- After prolonged use or if your hands begin to sweat

possible bacteria and the food. Obviously, if used incorrectly, gloves are no better than bare hands. Some studies have shown them to be worse; they can become contaminated and give a false sense of security. Gloves are made from many materials including latex, polyvinyl, nitrile, chloroprene, and polyethylene. Latex is the most problematic of the group because a small amount of the population is sensitive to it and trace amounts can find their way into the food. Gloves should be worn by anyone with cuts and open sores; a glove should be placed on an already bandaged hand. Because workers with an open cut can unknowingly infect others, they should be reassigned to duties that do not require contact with food and customers.

Excessive perspiration, which happens when gloves are worn for an extended period of time, is a breeding ground for bacteria especially when combined with the heat of the hands. In this situation, gloves should be changed as frequently as possible and hands should be thoroughly washed and dried between applications.

TEMPERATURE DANGER ZONE

Pathogens can live at all temperature ranges. Those capable of causing food-borne illness thrive in temperatures ranging from 41° to 135°F/5° to 57°C, the temperature danger zone. Most pathogens are either destroyed or will not reproduce at temperatures above 135°F/57°C. Storing foods below 41°F/5°C will slow or interrupt the cycle of reproduction.

Because pathogens can reproduce quickly, it is critical to reduce the time during which foods remain in the danger zone; those left in the zone for more than 4 hours are considered contaminated and should not be served. This 4-hour period is cumulative, meaning that every time the food enters the danger zone the meter is running. Once this 4-hour window has been exceeded, heating and cooling cannot recover the food.

Because improperly cooled foods are a leading cause of food-borne illness, cool all foods down as quickly as possible. Cooked foods that are to be stored need to be cooled quickly to below 41°F/5°C, unless using the two-stage method of cooling. First, foods must be cooled to 70°F/21°C within 2 hours. Second, foods must reach 41°F/5°C or below within an additional 4 hours, for a total cooling time of 6 hours.

PROPER COOLING PROCEDURES

- Hot liquids should be placed in a metal container and cooled in an ice water bath, stirring to bring down the temperature more rapidly
- Semisolid and solid foods should be refrigerated in a single layer in shallow containers

- Large cuts of seafood or meat should be cut into smaller portions, cooled to room temperature, and wrapped before refrigeration

PROPER REHEATING OF FOODS
- Move prepared foods quickly through the danger zone
- Reheat foods to 165°F/74°C for a minimum of 15 seconds

RECEIVING AND STORING SAFE FOOD
- Inspect all food for sanitary conditions
- Check the temperature of the food and the delivery truck
- Check the expiration date of the food
- Verify that foods have the required inspection certificates, or tags in the case of shellfish
- Transfer foods as quickly as possible to proper refrigeration

SYMPTOMS

Because most contaminants are ingested through the mouth, it is the gastrointestinal tract that presents the common symptoms of diarrhea, vomiting, and cramps, the culprit being a variety of microbes. It should not be assumed that the symptoms are caused by food. Water, ice, and hand-to-hand contamination are other ways that infections are spread.

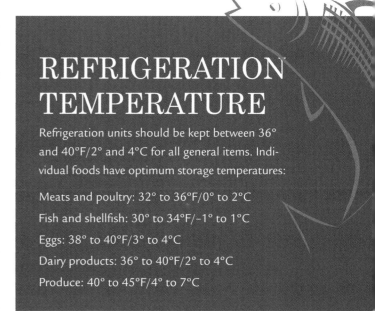

REFRIGERATION TEMPERATURE

Refrigeration units should be kept between 36° and 40°F/2° and 4°C for all general items. Individual foods have optimum storage temperatures:

Meats and poultry: 32° to 36°F/0° to 2°C
Fish and shellfish: 30° to 34°F/−1° to 1°C
Eggs: 38° to 40°F/3° to 4°C
Dairy products: 36° to 40°F/2° to 4°C
Produce: 40° to 45°F/4° to 7°C

COMMON FOOD-BORNE DISEASES

Salmonella
Salmonella is a bacterium spread by eating foods contaminated with the salmonella bacteria. Found in the intestines of birds and mammals, it is transmitted by eating undercooked poultry, eggs, mayonnaise, milk, and soft cheese. Studies have shown chickens and eggs may harbor salmonella on their skin or shells, even if they are "free range" or "organic."

Preventive steps include:
- Pay optimum attention to personal hygiene
- Use sanitizing solutions on work surfaces and cutting equipment
- Cook food to proper temperatures

E.coli 0157:H7
E.coli 0157:H7 is a bacterial pathogen found in the intestinal tract of mammals. Human cases typically are caused by ingesting contaminated food or water. Preventive steps include:

- Cook all ground beef until the juices run clear
- Consume only pasteurized milk and juice products
- Wash hands for 20 seconds with hot water and soap after using the toilet

- Sanitize work stations and equipment after handling raw meat
- Drink water that is treated by approved methods
- Avoid contact with persons infected with *E.coli* 0157:H7

Clostridium botulinum

Clostridium botulinum, also referred to as botulism, is a collection of bacteria typically found in dirt. They form spores that enable them to lie dormant until proper growth conditions exist. In the foodservice industry bacteria can be present in improperly canned or jarred foods, which are easily identified by bulging and swollen lids.

Anisakis

Anisakis are parasitic worms that live and hatch in the sea. Their larvae are consumed by crustaceans that are subsequently eaten by fish. The fish play host to the larvae, which in turn are consumed by humans who become infected with the disease. Consumption of raw seafood is the primary source of infection. These roundworm nematodes grow to about an inch and are as thick as the point of a pen. They are off-white or brown in color and will be straight or coiled. The worms can be identified by placing the fillet on a light box, and they are easily removed with a fillet knife and pliers. Symptoms of anisakis occur within an hour of eating infected seafood and include nausea, vomiting, and abdominal pain. More complicated symptoms can present if the larvae pass through to the bowel. The most efficient means of preventing infection is to cook raw fish to 140°F/60°C or freeze it to –4°F/–20°C.

Tapeworms

Destroyed by proper cooking or freezing, tapeworms are found in both fresh-water and saltwater fish. Their larvae look like a grain of rice and can be identified by using a light box. They can be easily removed from the fillet. Mild stomach pain and anemia can occur; diagnosis is usually made through a stool sample.

Gempylotoxin

Found in the oily fish family of Gempylidae, including escolar and mackerel, these toxins proliferate in the oil-fatty tissue of the fish. They cannot be digested and cause diarrhea. Limiting consumption to just a few ounces is the best defense.

Listeria

Listeria is an infection that results from consuming food contaminated with the bacteria. Although not commonly associated with fresh seafood, it has been linked to cold smoked fish and raw fish that have been stored too long. Also found in raw vegetables and unpasteurized milk, it is easily killed with heat and can be avoided with proper sanitation practices.

Ciguatera

Ciguatera is a toxin found in over 400 species of warm-water tropical-reef fish throughout the Caribbean and Pacific. Although harmless to fish, in humans it can cause gastrointestinal and neurological symptoms lasting for weeks or, in rare cases, years. Symptoms appear within 6 hours of consumption; there is no antidote or treatment for ciguatera poisoning. Care involves managing and supporting the body's recovery, not treating the poison. The toxin occurs naturally in marine organisms and is found in higher levels at the top of the food chain. Fish such as grouper, barracuda, and amberjack are just a few that have been associated with the toxin. Because ciguatera is resistant to heat and cold, it cannot be eliminated by cooking or freezing. Avoiding suspected fish species is the only sure way to prevent ciguatera poisoning.

Scombroid

Also called *histamine poisoning,* scombroid is a type of food poisoning caused by eating foods with elevated levels of histamine. This can occur during production in items such as Swiss cheese, or by bacterial spoilage in improperly handled seafood. Mahi mahi, amberjack, and tuna are species associated with scombroid. However, all foods that are subjected to certain conditions and amino acids are vulnerable.

Symptoms include headaches, rash, and decreased blood pressure, which can progress to gastrointestinal problems and hospitalization in high-risk groups.

To avoid scombroid poisoning, special care should be taken when handling susceptible fish. Elevated histamine levels can be generated in just 6 hours if the fish is not refrigerated or iced. Treatment is typically with antihistamines.

Red tide

Red tide is a marine occurrence that takes place when microscopic phytoplanktons fueled by the sun's rays multiply and thrive in the spring and summer nutrient-rich ocean environment. Not always red, the algae bloom forms a variety of colors and density. An array of coastal marine wildlife including birds, fish, and mammals succumb to the toxic algae. Filter-feeding mollusks are often unaffected by the toxins, but can be poisonous when eaten raw.

Shellfish poisoning

Shellfish poisoning comes from consuming shellfish infected by eating toxic planktonic algae, with raw shellfish responsible for the majority of all illnesses associated with seafood. Mollusks are stationary filter feeders more prone to consume and filter toxins than fin fish. Strict regulatory enforcement of fishing grounds and tagging keep most cases confined to recreational harvesters who are not aware of restrictions and warnings. There are four syndromes associated with shellfish poisoning:

- Paralytic
- Neurologic
- Diarrheal
- Amnesic

PARALYTIC SHELLFISH POISONING: As the name implies, paralytic shellfish poisoning is one of the most dangerous of all syndromes associated with seafood. Consuming shellfish that contain the toxin is rare, because commercial harvesters are regulated and are kept apprised of ocean conditions and possible red tide outbreaks. Symptoms include respiratory paralysis; the illness can be fatal. Harvesting shellfish with the proper permits in monitored waters is one way to avoid shellfish poisoning. Further steps include proper refrigeration and avoidance of any raw shellfish, especially by those who are pregnant or have a compromised immune system.

NEUROLOGIC SHELLFISH POISONING: Symptoms of neurologic shellfish poisoning include numbness of the mouth and extremities, diarrhea, and vomiting.

DIARRHEAL SHELLFISH POISONING: Not fatal, diarrheal shellfish poisoning symptoms appear within an hour of consuming infected shellfish; they include headache, vomiting, and diarrhea.

AMNESIC SHELLFISH POISONING: Amnesic shellfish poisoning affects the brain; it can cause memory loss and brain damage, as well as death. Cooking or freezing the shellfish does not diminish the effects of the toxin, which can be very serious.

CURING, BRINING, SMOKING, RAW, AND CAVIAR

Food preservation has been practiced in one form or another from time immemorial. The techniques used have not changed significantly since the days when a side of salmon was placed on a pole near the fire to dry and smoke. Our ancestors realized early on that seafood and meats spoiled quickly after the fish or animal was killed and devised methods to preserve it and make it safe to eat.

With ice unavailable, the fishermen of Peru carried limes on fishing expeditions because the juice effectively denatures the protein. The Norwegian, Portuguese, and Basque would set sail with barrels full of salt and return them full of bacalao (salt cod) which even today is preferred over fresh cod by many Spanish and Portuguese.

Other ways of preserving foods evolved. Simply drying foods in the sun worked for items low in protein, but by this method meat and fish would often spoil and rot. Salt, which was either evaporated from seawater or mined from the ground, was found to have a natural affinity both in taste and ability to draw out moisture. Salt alters foods by acting as a dehydrating agent, drawing out moisture through osmosis, dehydration, fermentation, and denaturing of proteins. It lowers the water content of bacterial organisms, which limits their ability to thrive and reproduce.

Smoke is another means of preservation, but alone it is insufficient because it does not penetrate the food. Many foods that are smoked are first cured or brined. Curing involves enrobing a piece of seafood in a mixture of salt, sugar, and seasonings for a specific amount of time based on its size, weight, and characteristics.

After the product is cured, it is allowed to dry uncovered to form a pellicle or skin on the outer part of the fish, which enables the smoke to adhere more consistently. Once the fish has cured and dried, it is ready for either the hot or cold smoker.

Brine is a mixture of water, salt, sweetener, spices, and sometimes chemicals such as sodium nitrite. Shellfish, pork, beef, and poultry benefit from being submerged in liquid for varying amounts of time. The water and salt penetrate the item, adding moisture and creating an inhospitable environment for bacteria. Brined foods have a pleasant seasoned taste and are typically juicy from the moisture that was absorbed. Some brined foods are smoked, but others are roasted, braised, or stewed. Brines are used by the food industry as a way to consistently deliver a specific flavor profile and moisture level, as well as better profit margins.

With new technology available worldwide, consuming foods preserved by ancient methods is a choice; we enjoy the taste and will continue to eat cured, pickled, and smoked seafood. Modern refrigeration has allowed us to discover and appreciate raw seafood. For generations, food was cooked to make it palatable, free from pathogens, and safe to eat. This drastic reversal is not for the squeamish; but with proper preparation and storage methods, eating raw seafood is perfectly safe, and allows for experimentation with new textures and flavors.

This chapter discusses the techniques and ingredients used to prepare many of these traditional culinary delicacies. Extreme care should be taken to ensure the freshest product available when serving cured, smoked, or raw seafood. The guidelines listed in this chapter are not meant for the commercial market but for individual applications. Commercial sale of smoked seafood products requires adhering to state and federal regulations, so check with your local health department for guidance if you are considering these items for your restaurant. Strict sanitation procedures should be in place, including a Hazard Analysis Critical Control Point (HACCP) plan, to ensure that safe food is served.

SALT: SODIUM CHLORIDE (NaCl)

Whether it has been mined from the ground or evaporated from the ocean, all salt is marine based and is composed of two minerals, sodium and chloride. Underground rock salt emanated from large lakes that dried up or had been evaporated from large ponds. Ancient windmills pump seawater into low-lying, often clay-bottom flats that are dammed off, leaving the wind and sun to evaporate the water. The residue is wet salt that can be white, brown, red (see Figs 11.1a and 11.1b), gray, or black, with varying subtle flavors depending on the surrounding environment or specific mineral content such as volcanic ash, clay, or even algae. Processing involves drying, sorting, and grinding, depending on the use, and can be valuable even today with scarce specialty varieties costing upwards of 100 dollars a pound.

FIGURE **11.1a** Red salt from the United States.

FIGURE **11.1b** Gray sea salt from France is evaporated from shallow clay bottom salt flats. Sun and wind evaporate the water leaving the salt, which is collected, dried, and graded for market. Different flavors and colors come from the minerals found in the sea water.

SALT IN THE KITCHEN

Most chefs prefer kosher or sea salt because of its ease of use and purity. Additives to common table salt include sodium alumino-silicate, or tricalcium phosphate, which are used as anti-caking agents. Others such as sodium iodide are used to cure iodine deficiency in humans, a leading cause of goiter. Kosher salt is usually pure salt (sodium chloride) of a large size, its surface area able to absorb moisture and better adhere to the food. Its absorbent qualities are important to the koshering process, which requires removing the animal's blood as efficiently as possible. Ordinary table salt is also kosher, but dissolves quicker and is therefore not as useful. Because kosher

SALT VARIETIES

All salt is not created equally and there are varieties available for all applications and price ranges.

- Popcorn salt
- Iodized salt
- Fluorinated salt
- Kosher salt
- Table salt
- Curing salt
- Pretzel salt
- Pickling salt
- Fleur de sel
- Japanese Nazuna sea salt
- Smoked sea salt
- Hawaiian red Alaea sea salt
- Andes Mountains pink salt
- Cyprus black lava flaked sea salt
- Indian Black sea salt
- Flavored salts
- Salt substitutes

is a pure salt, it does not have an iodine flavor which, when combined with seafood already high in iodine, would accentuate the metallic flavors. Kosher salt is also favored by most professionals because its larger size enables the user to more accurately judge the amount to be used.

Evaporated from many sources worldwide, sea salt is rich in flavor, color, and texture and is used by chefs as a finishing salt. However, there are exceptions. Using it to season a batch of clam chowder is not cost effective because the salt's unique qualities are lost once dissolved and mixed with other ingredients. The best way to bring out the subtle nuances of sea salt is to apply a small amount of it right before the dish is served. A pinch over perfectly seared scallops or a tossed salad noticeably highlights its flavor and texture.

Because some salt is large enough to break a tooth, it is important to grind it down to manageable sizes, which is easily done in a salt grinder or mortar and pestle.

HISTORY OF SALT

The importance of this simple nutrient was far reaching in the ancient world. References throughout the Bible and Greek mythology extol its special properties. The axiom "all roads lead to Rome" may refer in part to the Via Salaria (salt road) that extended from the Port of Salaria (salt port) just outside of Rome, east to the Adriatic Sea. Many scholars believe that Rome's location was selected because of its proximity to the abundant salt marshes of Ostia at the mouth of the Tiber River.

The ancients realized early on that salt extracted moisture and extended the life of high-protein items. Cheese was rolled in salt or submerged in brine to begin the drying process. Meats and fish were salted and hung and made into the modern equivalents of beef jerky and salted fish. Ground meats in the form of sausage were salted and dried making a shelf-stable, high-protein food suitable for long journeys. Garum, the famous sauce of the Roman Empire, was made by fermenting fish entrails in a brine solution that had flavor characteristics similar to modern-day Asian fish sauce. Other foods that were salted were mullet or tuna roe botarga, as well as vegetables such as peas, asparagus, mushrooms, and artichokes.

HEALTH

Salt contains sodium and chloride, which are essential to life; they regulate fluid balance in all living organisms. Since the 1950s, salt has been linked as a possible contributor to high blood pressure. Most recent research suggests that a high salt diet does not necessarily contribute to high blood pressure in healthy people but may have an adverse effect for those sensitive to sodium or for those with hypertension or kidney malfunction. The majority of sodium in the American diet is a result of pre-prepared and fast foods. Current regulations by the Food and Drug Administration (FDA) suggest limiting daily sodium consumption to 2,400 milligrams, which is equivalent to about one teaspoon of salt per day.

CURING

Salt has been used for centuries to inhibit bacterial growth and extend the shelf life of fish. Generally, there are two accepted methods of curing, depending on the fat content of the fish. Dry curing, the process of enrobing the fish in salt was, and still is, used for lean fish such as cod. Fish that are high in oils such as mackerel, tuna, and herring typically are brined, which is a process of curing in a salt/liquid solution. For centuries most fish not eaten immediately were salted. Sometime in the thirteenth century it was discovered that if these oily fish were placed in barrels and layered in salt, the salt would draw out the moisture from the fish and form what is now known as brine. The brine acted to equalize the osmotic balance of the fish and improve its longevity. Brining is important for species high in oil because the oils go rancid quickly, especially when surrounded by oxygen. Once a fish is submerged in a brine solution, in a sealed barrel, the oxidation is eliminated and the equalization properties of the salt can take place.

Another early discovery was a technique of gutting the fish. Early on, all fish were cured with the guts intact that inevitably added an interesting and perhaps unwanted flavor profile to the finished product. Eventually, a one-step technique of gutting the fish evolved. By cutting behind the gills in one motion, the heart, liver, stomach and some of the intestines and gill were extracted. What was left was the head and a part of the fish called the *pyloric caeca,* which are small sacs attached to the intestinal tract behind the stomach. These sacs contain an enzyme that breaks down the flesh, contributing an interesting fermented cheese–like flavor, which was new and unique at the time.

CURES AND BRINES

Cure is a generic term used for brines, pickling, corning solutions, and dry cures. Although unrefined salt and seawater were most likely the original cures or brines, we have learned over time how the individual components of cures and brines work. Refined and purified salts, sugar, spices, and sometimes chemicals (nitrates and nitrites) have made it possible to regulate the process more predictably. This means we can now produce high-quality, wholesome products with the best texture and taste.

DRY CURES

A dry cure can be as simple as salt alone, but more often it is a mixture of salt, sweetener, and spices. Keeping foods in direct contact with the cure helps ensure an evenly preserved product. Some foods are wrapped in cheesecloth or food-grade plastic; others are packed in bins or curing tubs with layers of cure packed between the products. Foods should be turned or rotated periodically as they cure. This process is known as overhauling.

FIGURE 11.2a Mise en place for gravlax cure is a neutral spirit (aquavit), chopped dill, brown sugar, and kosher salt.

FIGURE 11.2b Brush the aquavit onto the flesh of the fish. Mix the salt and sugar together and lay half of it on the cheesecloth. Place the skin side onto the cure.

FIGURE 11.2c Apply the rest of the cure to the flesh side of the fillet. Place the dill onto the top of the cure.

FIGURE 11.2d Wrap the fish in the cheesecloth. Place another pan or a weight on the fish and cure under refrigeration for 2 to 3 days.

FIGURE 11.2e After curing for 3 days, unwrap the salmon, brush off the cure with a damp cloth, and it is ready to slice and eat.

FIGURE 11.3 After curing the cod fillet in salt, remove it and keep under refrigeration until ready to use. Salted cod should be repeatedly soaked in fresh water to leach out the salt.

FIGURE 11.4a To prepare the cure for smoked salmon, combine two parts salt with one part sugar and mix in the spices.

FIGURE 11.4b Cover the bottom of a hotel pan with plastic wrap and add half the cure to the bottom of the pan.

FIGURE 11.4c Lay the skin side of the salmon fillet on the cure and cover the flesh side with the remaining cure.

FIGURE 11.4d Wrap the salmon with plastic and weigh the cure down using another hotel pan. Refrigerate approximately 20 hours.

BRINES

In today's kitchen, brining refers to the process of submerging a product in a mixture of water, salt, sweetener, and spices. Typically, this process is done to add flavor and moisture and to preserve foods that will be smoked. Kosher salt is a good choice for brining because of its purity. Avoid salt containing iodine because it could highlight the natural iodine in certain seafood and have a negative effect on taste.

SALT AND THE CURING PROCESS

Salt changes foods by drawing out moisture and other impurities, making it less susceptible to spoilage. This is accomplished through a basic process involving:

- Osmosis
- Dehydration
- Fermentation
- Denaturing of proteins

OSMOSIS

Osmosis in the curing process involves the withdrawal of water from the meat or fish to equalize the concentration of the salt. The fluid inside the cell membrane of the cured food travels through the membrane to dilute the salt on the other side. Once there is more fluid outside the cell than in, the fluid returns to the cell's interior, taking with it the dissolved salt. This process of flooding (equalizing) the inside of the cell with salt to kill the harmful pathogens is the essence of curing.

DEHYDRATION

Dehydration is the process of removing the moisture from food using heat, salt, or air for a controlled period to eliminate microbial growth. These methods can be used alone or in conjunction with each other to achieve a safe product.

FERMENTATION

Substances known as enzymes feed on the compounds found in energy-rich foods such as meats, fish, and grains. They ferment the food by converting the compounds in these foods into gases and organic compounds. The gases may be trapped, producing an effervescent quality in beverages, holes in cheeses, or they might disperse, leaving the organic acid, as occurs when preparing sauerkraut or pickles.

By increasing the acid levels in foods, enzymes also help preserve them, as most harmful pathogens can thrive only when the levels of acids are within a specific range (pH). Higher acid levels mean that the food's flavor is changed as well: it is sharper and more acidic.

Left unchecked, the process of fermentation would completely break down a food. Salt is an important control on this process, as it affects how much water is available to the enzymes. Like bacteria and other microbes, enzymes cannot live without water. Salt absorbs the water and prevents fermentation from getting out of control.

DENATURING PROTEINS

Whenever food is preserved, the structure of its proteins is changed. This change, known as denaturing the protein, involves the application of heat, acids, alkalis, or ultraviolet radiation. The strands that make up the food's protein lengthen or coil, open or close, recombine, or dissolve in such a way that foods that were once soft may become firm, smooth foods may become grainy, and translucent foods may become cloudy. Firm foods may soften and even become liquid. Examples of these changes include the denaturing of raw fish when making ceviche, or the blooming of gelatin.

CURING SALTS: NITRATES AND NITRITES

For millennia, meats and fish have been cured with unrefined salts and comprised an important part of the diet of our ancestors. Those meats took on a deep reddish color. It was not realized why until the twentieth century, when scientists unlocked the mystery of how nitrates and nitrites — compounds present in unrefined salts caused cured meats to redden. Saltpeter, or potassium nitrate, the first identifiable curing agent, did not produce consistent results; the color of the meat did not always set properly, and the amount of residual nitrates was unpredictable. Its use has been limited since 1975, when it was banned as a curing agent.

Nitrates (NO_3) take longer to break down in cured foods than nitrites do. Foods that undergo lengthy curing and drying periods must include the correct level of nitrates. Because nitrites (NO_2) break down faster, they are appropriate for use in any cured item that will later be fully cooked.

SWEETENERS

Salt-cured foods have a harsh flavor unless sweeteners are included in the cure. A range of sweeteners such as dextrose sugar, corn syrup, maple syrup, and honey can be used for their individual flavors and sweetening power. Honey added to brine will add an interesting light brown color to smoked foods due to its hydroscopic properties. Dextrose is often used commercially in cures; it mellows the harsh salt and increases moisture without adding an extremely sweet flavor of its own. Sweeteners can:

- Balance the overall flavor
- Overcome harshness of the salt
- Increase water retention in finished products
- Add to the overall color of the finished product

SPICES AND HERBS

A variety of spices and herbs are used in the curing and brining process to enhance a product's flavor and give it particular characteristics. Traditionally, many sweet spices such as cinnamon, allspice, nutmeg, mace, and cardamom have been used.

NITRITE AND NITRATE CONSUMPTION

Sodium nitrate and sodium nitrite are important elements in keeping smoked fish safe from botulism. When nitrates and nitrites break down in the presence of extreme heat, potentially dangerous substances known as nitrosamine may form in the food.

The presence of nitrosamines in cured products has been a concern since 1956, when they were discovered to be carcinogenic. Individual nitrosamine levels are influenced by the levels in food, as well as the amount produced by the salivary glands and in the intestinal tracts. Although more than 700 substances have been tested as possible nitrate replacements, none has been identified as effective. Nitrites do pose some serious health threats when they form nitrosamines, which may happen under extreme heat. There is little doubt that without nitrites, however, deaths from botulism would increase significantly and pose a more serious risk than the dangers associated with nitrosamines. The use of nitrates and nitrites is closely regulated and they should be used with extreme caution.

In addition, ingredients such as dry and fresh chilies, herb pastes, infusions, and essences, wines, fruit juices, and vinegars can be incorporated to give a contemporary appeal to cured seafood.

COMPARISON OF THE DIFFERENT CURING METHODS

DRY CURE	WET CURE (BRINE)
Used for products when a lower water content and dryer texture is required, such as smoked salmon	Used for items that will be hot-smoked or cooked
Longer shelf life for commercial products because of lower moisture content	Product will absorb moisture, increasing its weight and making it more difficult to caramelize. Some scallop brines contain chemicals such as sodium tripolyphosphate

METHODS FOR CURING SEAFOOD

DRY CURE	WET CURE (BRINE)
Fabricate product as necessary	Fabricate product
Prepare dry cure	Prepare brine
Calculate amount of dry cure to be used according to weight of product	Submerge in brine
Cover the entire product in cure	Place item in non-corrosive container, cover with brine
Wait required amount of time	Wait required amount of time
Rinse product	Rinse product
Air-dry to form a pellicle	Air-dry to form a pellicle
Hot or cold smoke product	Hot or cold smoke product

Note: Maintain proper sanitation practices throughout the entire process of preparation. Consume all cured, brined, or smoked products within a few days of preparation practicing a first in, first out (FIFO) procedure. Label and date all foods.

DRY CURING OF SEAFOOD

- Length of cure time is based on the water/salt phase
- Dry cure amount is not calculated because of shorter cure time

METHOD

1. Trim item.
2. Pack a liberal amount of prepared dry cure on item and place in a non-corrosive container.
3. When necessary amount of time has elapsed, rinse item, place on rack, and air-dry in refrigerator overnight.
4. Hot or cold smoke.

DRY CURE FORMULAS FOR SEAFOOD

INGREDIENTS	DRY CURE FOR COLD-SMOKED SALMON	DRY CURE FOR GRAVLAX
SALT	1 lb.	8 oz.
SUGAR	8 oz.	1 lb.
SEASONINGS	1 Tbsp. onion powder 1-½ tsp. ground cloves 1-½ tsp. ground bay leaf 1-½ tsp. ground mace ½ tsp. ground allspice	3 Tbsp. black pepper, cracked 2 bunches dill, chopped Juice of 2 lemons

Note: Check with your local health department for regulations regarding curing and cold smoking of seafood.

DRY CURING TABLE FOR SEAFOOD

ITEM	APPROXIMATE LENGTH OF TIME
Shrimp/scallops	½–1 hour
Trout fillet	2–4 hours
Whole trout	4–6 hours
Salmon fillet	8–20 hours for smoked salmon 24–72 hours for gravlax

BRINE FORMULA FOR SEAFOOD

INGREDIENTS	FISH
Water	2 gal.
Salt	3-1/2 lb.
Sweetener	1 lb.
Seasoning	1 cup pickling spice 2 cloves garlic 1/3 cup lemon juice

Note: Amount of sweetener depends upon the type used (degree of sweetness).
Method: Combine all ingredients in a non-corrosive container and mix until dissolved.

WHAT CAN GO WRONG WITH BRINED AND CURED ITEMS

PROBLEM	SOLUTION
Item is too salty or not salty enough	Cured too long or not long enough Recipe ingredients are incorrect
Gray spot or area Item is not completely cured to the center (if nitrites are used, this will not have the pink color and will still be the natural meat color)	Cure or brine time is not long enough Area for curing is too cold In most cases, problem is solved by extending curing time
Brine or item spoils quickly	Area for curing is too warm Recipe ingredients are incorrect In most cases, this is a problem caused by insufficient salt or improper length of cure time

Note: Always brine and cure products under proper refrigeration.

SMOKE

Smoke has been applied to food since it was first recognized that holding products high in protein near a fire not only dried them quickly or prevented animals and insects from invading them, it produced a special flavor that people have enjoyed for centuries. Today, many smoked products are featured on the menus of all manner of restaurants.

Smoke is varying shades of white. Under a microscope, it might be seen to contain two parts, vapor and particles. Recognizing its composition, one has a better understanding of how foods are preserved through the smoking process. Traditionally, the product is surrounded by smoke, a complex mixture of many chemicals and compounds that occur when wood and other items are burned. It is the smoldering and smoking of the wood that creates smoke, not the flames. Actual fire in the smokehouse is not recommended, because a controlled heat and smoke environment is important to the final outcome.

FORMATION OF THE PELLICLE

Smoke sticks to food most efficiently if the food is first dried. Product should be allowed to air-dry long enough to form a tacky skin known as a pellicle. The pellicle plays a key role in producing excellent smoked items. It acts as a protective barrier for the food and helps capture the smoke's flavor and color. Seafood can be properly dried by placing it uncovered on racks in the refrigerator. To encourage the formation of the pellicle, place the foods in an area that has good air circulation. The exterior of the item must be sufficiently dry if the smoke is to adhere to it.

WOODS AND OTHER ITEMS USED FOR SMOKING

Smoky fires are ignited in special smoke chambers to control:

- Oxygen — a decrease in oxygen causes wood to smolder and smoke
- Moisture — damp products smolder rather than burn

WOODS FOR SMOKING

Hard, fruit, or nut woods are preferred. All woods impart their own distinctive flavor (see the following chart). Wood is available in sawdust, chip/nugget, and chunk form (see Fig 11.5). Each smokehouse accepts different forms of wood, so check with the manufacturer for recommendations.

Soft or resinous woods should not be used because they will flare up, burn, and impart a bitter taste. Woods should be free of contaminants. Never use pressure-treated wood, which is toxic when burned.

MATCHING SMOKING WOODS WITH FOODS

WOOD TYPE	CHARACTERISTICS	GOOD FOOD MATCHES
Hickory	Pungent, smoky, bacon-like flavor.	Pork, chicken, beef, wild game, cheeses
Pecan	Rich and more subtle than hickory, but similar in taste. Burns cool, so ideal for very low heat smoking.	Pork, chicken, lamb, fish, cheeses
Mesquite	Sweeter, more delicate flavor than hickory. Tends to burn hot, so use carefully.	Most meats, especially beef. Most vegetables
Alder	Delicate flavor that enhances lighter meats.	Salmon, swordfish, sturgeon, other fish. Also good with chicken and pork
Oak	Forthright but pleasant flavor. Blends well with a variety of textures and flavors.	Beef (particularly brisket), poultry, pork
Maple	Mildly smoky, somewhat sweet flavor. Try mixing maple with corncobs for ham or bacon.	Poultry, vegetables, ham, bacon
Cherry	Slightly sweet, fruity smoke flavor.	Poultry, game birds, pork
Apple	Slightly sweet but denser, fruity smoke flavor.	Beef, poultry, game birds, pork (particularly ham)
Peach or Pear	Slightly sweet, woodsy flavor.	Poultry, game birds, pork
Grape vines	Aromatic, similar to fruit woods.	Turkey, chicken, beef
Wine barrel chips	Wine and oak flavors. A flavorful novelty that smells wonderful, too.	Beef, turkey, chicken, cheeses
Seaweed	Tangy and smoky flavors. (Wash and dry in sun before use.)	Lobster, crab, shrimp, mussels, clams
Herbs & Spices (bay leaves, rosemary, garlic, mint, orange or lemon peels, whole nutmeg, cinnamon sticks, and others)	Vary from spicy (bay leaves or garlic) to sweet (other seasonings), delicate to mild. Generally, herbs and spices with higher oil content will provide stronger flavoring. Soak branches and stems in water before adding to fire. They burn quickly, so you may need to replenish often.	Vegetables, cheeses, and a variety of small pieces of meat (lighter and thin-cut meats, fish steaks and fillets, and kebabs)

Source: © Weber

FIGURE 11.5 Various wood forms can be used when smoking fish. Size and shape depend on the type of smoker being used. Pictured clockwise from left: wood chips, wood chunks, compressed sawdust disks, sawdust.

Reasons why seafood is smoked
Aroma and flavor
- Several hundred compounds including phenols and organic acids contribute to the individual smoke taste.

Preservation
- Compounds in the smoke have antimicrobial properties that affect only the outer surface of the product.

Color
- Carbonyl compounds in the smoke combine with the fish protein to form furfural compounds that give smoked food its traditional brown color.

Controlling oxidation
- Lipid oxidation which can degrade the flavor, texture, and color of the product is controlled by the smoke's antioxidant properties.

COLD SMOKING

Many foods are suitable for cold smoking. The basic criteria should be taste and texture. When dealing with animal proteins high in moisture, this is controlled with proper brining or curing times. In addition to fish, shellfish, and meats, cheese, beans, vegetables, and even tofu are commonly cold smoked. It is possible to use smoke as a seasoning by first smoking, then cooking the food.

Smokehouse temperatures for cold smoking should be maintained below 100°F/38°C. Some commercial processors maintain the temperature below 40°F/4°C to keep food out of the danger zone, especially when prolonged smoke times are required.

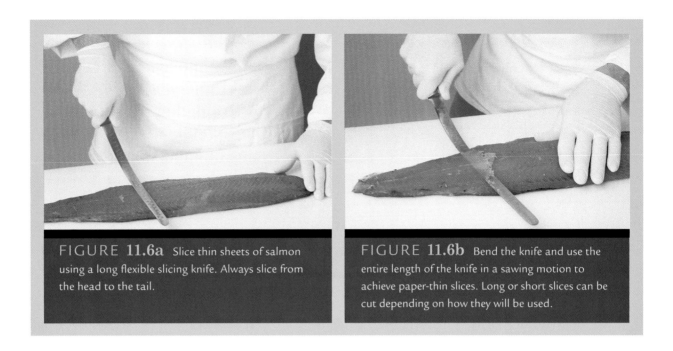

FIGURE 11.6a Slice thin sheets of salmon using a long flexible slicing knife. Always slice from the head to the tail.

FIGURE 11.6b Bend the knife and use the entire length of the knife in a sawing motion to achieve paper-thin slices. Long or short slices can be cut depending on how they will be used.

Keeping the smokehouse temperature below 100°F/38°C prevents the protein structure from denaturing. A cured and cold-smoked side of salmon is an attractive addition to a buffet table and the process it has undergone allows it to be sliced extremely thin.

HOT SMOKING

Hot smoking exposes food to smoke and heat in a controlled environment allowing it to cook and smoke at the same time. Hot smoking occurs within the range of 165° to 185°F/74° to 85°C which ensures foods are fully cooked, moist, and flavorful. A hotter temperature will shrink and dry the item.

SMOKING METHODS		
	COLD SMOKING	**HOT SMOKING**
Temperature of Smokehouse	70°F/21°C and 100°F/38°C (80°F/27°C is average temperature) Some commercial processors smoke fish at 40°F/4°C	160° to 185°F/74° to 85°C
Result of Smoking	Product does not cook Slight dehydration of overall texture	Product cooks during the smoking process
Storage	Refrigerate immediately after smoking	Final internal temperature of cured hot smoked seafood products should be 145°F/63°C. Refrigerate or serve immediately.

PROCESSING ANALYSIS FOR HOT-SMOKED AND COLD-SMOKED FISH

Receiving
- Fresh seafood must be received on ice at a minimum temperature of 41°F/5°C, have clear eyes, bright gills, and smell clean without a strong fishy odor.
- Because the freezing process alters the salmon's texture, only fresh fish is recommended for cold smoking.

Evisceration
- Fish must be eviscerated in a separate area and completely washed before curing.

Brining and curing
- Do not mix fish species in a brine or cure.
- All brining and curing should take place under refrigeration.
- Thickest piece of fish must have a final water phase salt (WPS) of 3.5 percent.
- Brines and cures should be used only once.

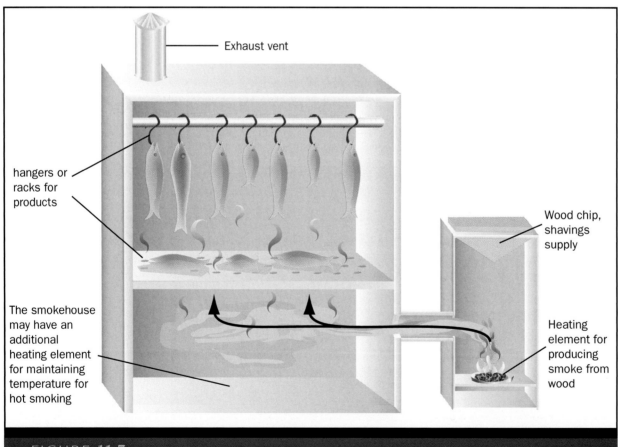

FIGURE 11.7a **Cold smoking** requires lower temperatures and usually does not need an additional heat source **Hot smoking** requires additional heat source to raise the temperature of the smokehouse to allow products to cook during the smoking process

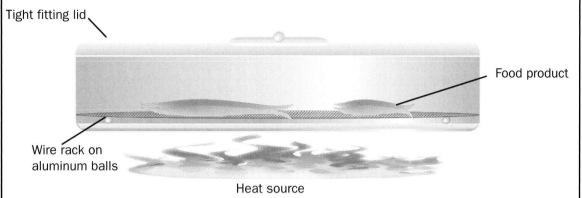

FIGURE 11.7b

Hot smoking
- To facilitate complete and even smoking, fish should not be crowded in the smoker.
- Temperature of the smoker must reach at least 145°F/63°C for at least 30 minutes.
- Temperature from the thickest part of at least three fish must be taken, with the lowest temperature recorded on the process record. Temperatures should be taken at least three times during smoking.

Cold smoking
- To facilitate complete and even smoking, fish should not be crowded in the smoker.
- Smokehouse temperatures should not exceed 50°F/10°C for more than 24 hours or should not exceed 90°F/32°C for more than 20 hours.
- Smokehouse temperatures should be taken at least three times during smoking.

Cooling
- Fish should be cooled to 40°F/4°C as quickly as possible and maintained at this temperature until sold.

Packaging
- For commercial sale, it should be sold air-packaged and labeled "Keep refrigerated at 38°F or below."

Recordkeeping

Detailed records must be kept on each batch of seafood and include the following detailed information:

- Product name
- Lot code
- Date processed
- Container size, number of containers (if applicable)
- Temperature record of thawing, brining, smoking, cooling, and storage of each batch of fish processed
- Records should also detail the smoke time
- Initial sale information pertaining to addresses and lot codes needs to be included

CAVIAR

Caviar is the salted roe of fish, traditionally sturgeon, which is prized for its briny delicate flavor. Due to overfishing, politics, and pollution, the supplies have dwindled and the price has increased, leading to a flood of alternative varieties from many sources.

The delicate berries of true caviar come from a variety of sturgeon, an ancient fish with a life span of over a century. The female has the ability to retain her eggs for several years waiting for the perfect conditions to continue the cycle: Because the sturgeon is a bottom dweller, it has encountered difficulty traversing man's obstacles along world rivers like the Thames, the Seine, and the Po, to name a few. Pollution in parts of the Caspian Sea has diminished the population, and longline fishing and countless ghost nets have led to a near extinction of the Beluga. Political tensions in Iran and the breakup of the Soviet Union, both of which control a majority of the Caspian's coastline, have also contributed to deregulation of both the fishing and processing industry. Azerbaijan, Kazakhstan, and Turkmenistan also border the Caspian Sea, an area that has for centuries had the perfect ecosystem and habitat for sturgeon. In a very short time, irrigation and industry has left the northern Volga delta changed forever, eliminating the world's best sturgeon spawning grounds.

CONSERVATION

Having roamed the world's waters since prehistoric times, sturgeon have both nourished and been highly prized by many ancient civilizations. If current practices continue, this medieval fish and its magnificent eggs will be lost to future generations. Scientists believe that banning all fishing in the Caspian Sea and regulating it in the rivers is the only way to ensure its survival. Many species of sturgeon are considered endangered, and legislation has been passed to make permits and customs declarations mandatory for importation of caviar. Any hope of saving the sturgeon population in the Caspian Sea may come from the many hatcheries worldwide that are working to raise and introduce millions of baby sturgeons or "fry" into the sea. In addition, many countries have thriving sturgeon farms that process both meat and eggs. Unfortunately, it is difficult and costly to differentiate gender, making release programs for female fish difficult. Sturgeons are such strong fish that eggs are harvested from mature females via caesarian section, after which they are sewn up and returned to the water. Sturgeons thrive in the warm water and controlled conditions of the farms, allowing them to mature and reach harvesting size quicker than in the wild. Meat from this boneless fish is low in fat and high in protein, and lends itself well to curing, drying, and smoking.

Although true caviar is said to come only from the Caspian Sea, sturgeon is found throughout the waters of Europe, Asia, and North America. Commercially, there are three varieties in U.S. waters: White sturgeon, Atlantic sturgeon, and the Great Lakes sturgeon. Because of overfishing and disruption of migration and spawning habitat, many state laws prohibit the catching of certain species and the selling of sturgeon eggs.

PROCESSING

As with all perishable products, speed is essential to freshness. Traditionally, sturgeons are netted from small boats by fishermen with vast experience. Once the fish are brought to the processing facility, they are cleaned and the entire membrane sac, including the eggs, is removed. The roe is then graded based on size, maturity, and color. To separate the eggs from the membrane, they are pushed through a sieve, then rinsed and cleaned. The washed roe is gathered up in small batches and carefully salted and stirred, a procedure which imparts to caviar its unique characteristics and flavor.

VARIETIES

Supported by a cartilaginous skeleton, sturgeons have no bones; they have not evolved since the Jurassic period. Although sturgeon easily crossbreed, Beluga, Sevruga, and Ossetra are the three varieties that are important for their caviar (see Fig 11.8). In the United States and France, only the processed roe of sturgeon is allowed to be called true caviar. However, the law is hard to enforce and has been interpreted more leniently in the United States, and as long as the type of fish is identified on the label, various kinds of fish roe may be sold as caviar.

GRADING

Individual caviar eggs are referred to as berries, which are graded for color, using a scale of 000 for the lightest colored caviar to 0 for the darkest. In addition to color and

FIGURE 11.8 From left to right: Ossetra caviar, Sevruga caviar, and Iranian caviar.

FIGURE 11.9 Unique because of irregular-shaped and colored eggs, Ossetra can be yellowish, olive green to brown with a distinct strong nutty flavor that mellows as the fish age. It is always sold in yellow jars or tins.

size, it is also graded within each variety. Beluga caviar has the largest eggs, and they are graded large or coarse. Ossetra eggs (see Fig 11.9) are slightly smaller than Beluga, and Sevruga are the smallest of all.

Beluga eggs, from the largest of all sturgeon, comprise over a quarter of its body weight and are pale to dark grey, almost black, with a delicate outer skin and a briny flavor. It has always been sold in blue cans or glass jars with blue lids. In 2005 the U.S. Fish and Wildlife Service banned the importation of beluga caviar from the Black Sea. Over the past several decades the population of beluga sturgeon has diminished to a critical point and for the species to survive the fish and its roe must be left alone.

Sevruga eggs are grayish green to black, fine grained, and have a unique and distinctive flavor. Because sevruga is the most plentiful, its caviar commands the lowest price. It is always sold in red jars or tins.

PRESSED CAVIAR

Pressed caviar is made from mature, broken, or overripe eggs. The salted eggs are put into a linen sack and pressed until all of the fatty liquid is released. The result is a marmalade-like substance that is often spread on black bread. The flavor of pressed caviar is more intense than other varieties.

OTHER VARIETIES OF "CAVIAR"

Other fish roes can be processed in the same manner as sturgeon roe and are available fresh and pasteurized. Roe from the salmon, paddlefish, crab, lumpfish, sea urchin, mullet, tuna, whitefish, flying fish, and cod each have unique sizes, textures, and flavors, and are sustainable and reasonably priced (see Figs 11.10 and 11.11).

FIGURE 11.10 From left to right: Whitefish caviar, Salmon caviar, Hackleback caviar, and Trout caviar.

FIGURE 11.11 From left to right: Tobyco Red, Tobyco Orange, Tobyco Black, Tobyco Golden, and Tobyco Green.

- Salmon caviar is golden pink to orange in color and the eggs are large.
- Whitefish from the Great Lakes and Canada produces a roe with a golden color and mild taste.
- Cod or carp roe is small and often salted and pressed.
- Tuna and mullet roe are common in the Mediterranean region where it is referred to as *boutargue* or *botarga*.
- Flying fish roe, known as *tobiko* are small in size and often colored with wasabi or other ingredients to produce unique colors and flavors.
- Sea urchin roe, known as *uni* in Japanese, is eaten both raw and processed and has a very fine texture and earthy flavor.
- Lumpfish roe is off-gray in color and is typically dyed red, gold, or black using vegetable-based dyes or cuttlefish ink. It is the least expensive of all roe available on the market.

QUALITY CHARACTERISTICS

Fresh sturgeon caviar should be plump and moist, with each individual egg shiny, smooth, separate, and intact. When you taste caviar, the eggs should release a savory flavor, faintly nutty with a hint of the sea, and they should snap in the mouth. It should be surrounded by its own thick natural oils but never appear to be swimming in it. If purchasing caviar in cans, the sides of the caviar should not fall down once the lid is removed. Caviar should always be received refrigerated.

STORAGE

It is imperative that from packing to purchasing, the product stay as close to 28°F/−2°C as possible. Once received, the caviar can be stored unopened for several months at slightly higher temperatures that should never exceed 37°F/3°C. At the processing facility, caviar is packaged in 4-pound tins and sealed with a wide rubber band. In Iran, each tin is labeled with detailed information including the variety of fish, grading, and weight of the fish's roe. These large tins are shipped worldwide where they are often repackaged in small glass jars weighing from 1 to 2-1/2 ounces (28 to 71 grams). Larger quantities are generally packed in tins with colored metal lids and are available in various weights from 4 ounces to 2.2 pounds (113 to 998 grams). At the very least, the jars should include an expiration date. When purchasing larger quantities, considerably more information should be made available. Handle the containers as little as possible; avoid exposing them to air and heat, which will greatly diminish the shelf life. Freezing is not recommended; it would deteriorate the delicate skin and ruin the texture and consistency. Because of the salting process, even at temperatures slightly below freezing, 27°F/−3°C, the caviar does not freeze. Once opened, it should be served immediately. Purchase only what you will consume in a day or two and avoid overhandling.

FIGURE **11.12a** Caviar should be simply served on ice.

FIGURE **11.12b** Serve caviar on materials that do not interact with the flavor. Pearl (pictured), bone, glass, wood, and even gold spoons are often used.

Pressed caviar will keep at 40°F/4°C for 10 to 14 days; once opened, it will last up to 5 days. The primary reason caviar goes bad after opening is cross-contaminations. Be sure to use a clean spoon and food-handling gloves when dispensing caviar.

Pasteurized roe and those processed with greater amounts of salt will keep in the refrigerator unopened for several weeks. Once opened, they should be consumed within several days.

CAVIAR SERVICE

True caviar needs no special accompaniments and its unique flavor should be served only with a palate cleanser such as Champagne or iced vodka. It is often served in special iced containers with mother-of-pearl, bone, horn, or glass spoons to avoid the flavor changes that may occur with certain metal spoons (see Figs 11.12a and 11.12b.b). Because it is so delicate and expensive, caviar should never be added to foods while they are cooking. If accompaniments are needed, then keep it simple. Crusty bread, crackers, or blinis can be used.

RAW

Many people indigenous to North America, including the Eskimos, have a rich culinary history of consuming raw seafood. For centuries, they have sustained themselves with a diet high in omega-3 fatty acids. Today, with the increased popularity of Japanese cuisine, many varieties of fish and shellfish are consumed raw.

Whenever foods high in moisture and protein are eaten raw, there are health concerns. Understanding freshness characteristics and educating yourself as to which species contain harmful parasites is important when deciding whether to eat raw fish and shellfish.

Worldwide there are also many varieties of seafood eaten raw, or cured in such a way that denatures the proteins, making them safe to eat.

- Carpaccio
- Tuna tartare
- Cured and smoked salmon
- Poke — Hawaiian raw seafood salad
- Ceviche
- Anchovies and herring, salted or brined

SASHIMI

Sashimi refers to a traditional Japanese preparation of excellent quality raw fish. Tuna, typically from the red meat species, is one of the most popular varieties, but a wide range of saltwater

> **LESS IS MORE**
>
> One of the most important aspects of recreating specific dishes from Japanese cuisine is to understand the important characteristics of each ingredient. The focus of these preparations should be on fresh taste and a composition of texture, color, and seasonality. Chefs train long and hard to master classical cooking techniques, methods, and combinations that involve many steps and ingredients that are married for the perfect dish. Sushi, and Japanese cuisine in general, is simpler in overall development and is characterized by the food's natural taste and appearance. The aim is to enhance the essential qualities of what is being served. Very few items in the sushi kitchen are meant to be prepared ahead of time. Sushi rolls must be made to order or the nori seaweed will lose its snap, the rice will become dry, and the delicate flavor of the raw fish will dry out and lose its luster.

species is suitable for this preparation. *Sashimi,* which in Japanese means "pierced flesh," is widely misunderstood by purveyors, chefs, and consumers. Cutting technique, species variety, and accompaniments, heretofore all foreign to most of us who grew up eating only cooked seafood, are being refined as sashimi increases in popularity.

TUNA

Tuna is sought-after for sashimi because of its wonderful flavor and texture (see Fig 11.13). Many people liken it to a perfect slice of cold rare prime beef. However the palate perceives it, its quality and fat content is dependent on its size, age, and species. Those perfect fish with the highest amount of fat command the highest price. Fishing methods, handling techniques, and chilling procedures are all factors that contribute to quality sashimi.

FIGURE 11.13 Tuna loin used for sashimi should be free of any blood line and be impeccably fresh.

Because the overall appearance of tuna is important to its value and quality, it should be landed alive and as rapidly as possible. When bringing the fish on the boat, it should be carefully gaffed in the head, not the body. Onboard, great care should be taken to avoid bruising any part of the fish. The brain, located behind the eyes, should be pierced with a spike. Oftentimes this is followed by the Taniguchi method, which involves inserting a length of ridged nylon or stainless steel through the brain and down the neural canal, destroying the spinal cord. This stops the chemical reaction of flesh deterioration and ensures higher quality. Proper bleeding is the next essential step, a procedure done to improve the appearance and extend the shelf life of the flesh for the sashimi market. Because internal organs such as intestines, kidneys, and gills contain large amounts of bacteria, these should be completely removed, followed by a thorough rinsing. Once these steps are completed, the fish is ready for onboard storage.

Tuna are warm-blooded (endothermic) in that they control their internal body temperature, which can rise to 86°F/30°C under certain conditions such as feeding or capture. It is imperative to lower the core temperature to 32°F/0°C as rapidly as possible and maintain it thereafter. Immediately after the initial landing and gutting procedures are completed, fish should be placed in a slurry of flaked ice and seawater, removed after 24 hours, transferred to a fish hold, and covered in ice. Although these procedures and methods seem extreme, they are important to ensure ultra-high qualities of sashimi. The USDA uses a numbering system of 1, 2, and 3 to distinguish quality. Within Japan there are additional criteria, all of which reflect the detailed care taken once the fish is landed.

Bigeye, yellowfin, and bluefin tuna are typically of sashimi quality. Evaluating the meat quality is done by cutting off the tails and examining the exposed flesh color and fat content. The pinkest, fattiest, and most valuable part of the fish is located on the underside near the belly and is called *toro*. This will have very fine white marbling similar to beef. The leaner sections close to the upper side of the backbone are called *akami*. Each of these two sections is graded high, medium, or low grade.

Much of the tuna found in the western Atlantic off the coast of New England originates in South America. As they migrate north through Florida, they spawn and are very lean. Once they reach their summer feeding grounds in New England, they fatten up on squid and herring before the fishing season opens in the fall. Some of these fish will be caught and flown directly to Tokyo's tsukiji market where they are graded, auctioned, and shipped to sushi restaurants worldwide.

SUSHI

The term *sushi* has evolved from its humble beginnings meaning "snack" in Japanese to a much broader term encompassing a wide range of preparations. Many countries throughout Asia, particularly Japan, sustain themselves with protein from the sea and carbohydrates from the rice fields. The relationship between these two foods is central to many aspects of culture and life. Ancient preservation methods involved pressing together rice (which was discarded after fermentation), salt, and raw fish. The lactic acid that formed prevented the fish from spoiling. This new flavor profile, although very strong, was pleasing for its ability to contrast the fish flavors with the ever-present accompaniment of neutral rice.

In the eighteenth century, this fermentation process was gradually eliminated and replaced with plain or seasoned rice and raw fish accompanied by variations in flavor, shape, and preparation of ingredients. Many modern interpretations are made with vegetables, beef, pork, chicken, and shellfish.

IT'S ALL ABOUT THE RICE

Sushi, in its simplest form, is a hand-manipulated portion of vinegar-seasoned rice.

Like many recipes that start with only two ingredients, rice can be a challenge. Using a rice cooker will greatly improve the quality and consistency of the finished product. Always use short grain sushi rice that typically calls for a ratio of 1 part rice to 1 part water by volume. Because the bland flavor of the rice is important to flavor contrast, stocks should not be used. Rice used for sushi is prepared in a two-step process of cooking and seasoning.

Cooking the sushi rice
- Measure the rice and place it in a bowl.
- Pour cold water over the rice and stir with your hands to remove the starch.
- Continue adding water until it runs clear, drain rice in a colander, and let rest 20 minutes.
- Place rice in the rice cooker; for each cup of dry rice, add 1 cup of water and 1 teaspoon of mirin; let this soak for 20 minutes.
- Cover and turn the rice cooker on. It will turn itself off in approximately 20 minutes.

Cooling and seasoning the rice
- Prepare a seasoning mixture of 1 part rice vinegar to 1/4 part sugar by volume.
- Once the rice is cooked, add a pinch of salt per cup of cooked rice and transfer it to a dampened wooden sushi bowl for cooling.
- Begin to cut through the rice with a wooden rice paddle as you pour the seasoning mixture over the rice.
- Fan the rice while carefully turning and folding to cool it and to allow the rice to absorb the seasoning and give it a shiny appearance.
- Stop when the rice is cool. Cover with a damp towel; do not refrigerate.

QUALITY CHARACTERISTICS OF SUSHI RICE
- Delicate balance of flavor between the rice, sugar, and vinegar
- Firm and sticky, not mushy and pasty
- Shiny and glossy, not dull and starchy

WRAPPINGS

FIGURE 11.14 Nori seaweed has a shiny and a dull side. The shiny side should be placed face down onto the bamboo mat so that it is on the outside of the finished roll.

Nori seaweed is used to roll up the sushi; it is available in square sheets from specialty markets. The shiny presentation side should be face down (see Fig 11.14). Commercially available sheets of processed and toasted seaweed are very nutritious. Because texture is very important, use only fresh crisp nori sheets and do not store them in the refrigerator. Nori is also used in soups and stews.

Rice paper sheets are made from a thin rice batter formed like crepes and cooked over steam. They should be briefly soaked in water or layered between damp cloths to rehydrate.

Japanese omelet
Called *tamagoyaki* in Japanese, this egg dish is cooked like a crepe, then rolled up and sliced or used as a wrapper. It is seasoned with mirin, soy sauce, and sugar.

Bean curd
These thin sheets of bean curd are typically eaten steamed or fried.

KEY INGREDIENTS
Many of the traditional ingredients in sushi have antibacterial properties that make it a healthy and safe food.

Vinegar
A French term meaning "sour wine," vinegar in the Japanese kitchen is made from rice and contrasts the neutral flavor of the rice that is served with sushi and sashimi.

Mirin
Mirin is a low-alcohol sweet wine that acts to contrast the vinegar in the sushi rice.

Soy sauce
Soy sauce is a very important component that harmonizes with all sushi ingredients. It contributes salinity, balances raw fish flavors, and contrasts the sweet and sour notes. The majority of soy sauce is dark; light and flavored varieties are also available. Tamari is a Japanese soy sauce product produced without the addition of wheat. Choose soy sauce that has a pleasant aroma and is dark but somewhat translucent.

Ginger
Ginger root should be firm and bright with moist-looking skin that can be easily removed with the back of a spoon. Ginger can be sliced, minced, or grated on a special board or mortar and is often pickled and served with sushi.

Wasabi — horseradish
Wasabi is a spicy radish that is grated into a paste and served as an accompaniment to sashimi and sushi. The traditional root is difficult to grow and most tubes and cans of wasabi are actually another variety of horseradish with the addition of colorings and mustard seed. Powder wasabi should be combined with equal parts water to produce a paste which can be formed into a small cone shape and served with soy sauce.

Vegetable
Sushi is impressive for its versatility of ingredients. Delicious flavors and textures can be achieved without animal proteins. Vegetables should be impeccably fresh and served at their height of ripeness. Knife cuts are extremely important to the presentation, so good-quality, sharp knives are essential to achieve precision

VEGETABLE VARIETIES FOR SUSHI
- Asparagus
- Avocado
- Burdock
- Carrots
- Cucumber
- Nattō
- Pickled vegetables
- Radish
- Snap peas
- Snow peas
- Certain varieties of squash and zucchini
- Soy beans

FINFISH VARIETIES

- Amberjack
- Bass
- Bigeye tuna
- Blue fish
- Bluefin tuna
- Bonito
- Cod
- Eel
- Flounder
- Halibut
- Mackerel
- Marlin
- Monkfish liver
- Sea bream
- Shad
- Snapper
- Swordfish
- Trout
- Yellowfin tuna
- Yellowtail

SHELLFISH, CEPHALOPODS, AND OTHER SEAFOOD

- Abalone
- Clams
- Crabs
- Lobster
- Mussels
- Octopus
- Salmon roe
- Scallops
- Sea cucumber
- Sea urchin roe
- Shrimp
- Squid
- Surimi (imitation crab meat)
- Whelks

cuts. Harmonize flavors, colors, and texture, along with good craftsmanship, for best results.

Seafood

In today's global world, a wide variety of seafood is available for the production of sushi and sashimi, with over half previously frozen. Advances in freezing technology allow for a superior product that maintains its structure, flavor, color, and fat content. Depending on supply and demand, seasonality of product, and final location, freezing can actually be beneficial. Salmon, known to have parasites, should not be eaten raw and should be frozen before being served sashimi style. Freshwater fish also have parasites and, if used, are first cooked.

When using fresh fish, standard quality characteristics are important to the final outcome. Clear eyes, bright gills, fresh clean aroma, and firm flesh are all indicators of freshness. Tuna is available whole, but commonly sold in loin form. Loins should be moist with a tight grain, fresh aroma, and bright color. Avoid loins that are dark and oxidized. Sushi-grade tuna is often misleading and misrepresented by purveyors. Tuna is one of the most sterile fish species and is safe to eat raw. Sushi grade is a quality reference to the various sections or cuts of the tuna based on the meat's color, texture, and fat content, not on whether it has been previously frozen.

SUSHI VARIETIES AND SHAPES
Nigiri sushi

FIGURE 11.15a Slice a 1-inch thick piece from the tuna and using your four fingers as a guide, cut through the slice. This will give you the appropriate length for your finished slice.

FIGURE 11.15b Slice through the block at an angle into ¼-inch strips.

FIGURE 11.15c Slice the rest of the block in the same way; the first and last slice, which will be triangular, can be sliced into two pieces.

FIGURE 11.15d Wet your hands in vinegar water so the rice does not stick to them.

FIGURE 11.15e Gather in your hand a mound of rice approximately the size of a golf ball.

FIGURE 11.15f Carefully squeeze the rice forming a length that is just a little smaller than your four fingers.

(continues)

(continued)

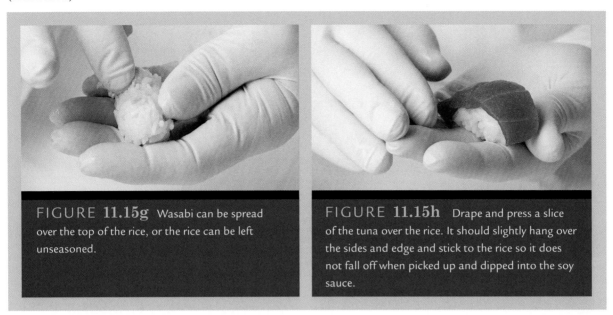

FIGURE **11.15g** Wasabi can be spread over the top of the rice, or the rice can be left unseasoned.

FIGURE **11.15h** Drape and press a slice of the tuna over the rice. It should slightly hang over the sides and edge and stick to the rice so it does not fall off when picked up and dipped into the soy sauce.

Nigiri refers to the technique of squeezing the rice into various shapes (rectangle, boat, fan, combs, balls) less than 3 inches in length and weighing about an ounce (see Fig 11.15e). Important to nigiri is the balance between the rice and the topping, so use only about 1/2 ounce of seafood over a small amount of wasabi (see Figs 11.15g and 11.15h). Avoid freshwater fish as they may be parasitic. Fish fillets or loins should first be cut into manageable rectangular pieces, then cut into diagonal strips about 5/16-inch thick and less than 3 inches in length.

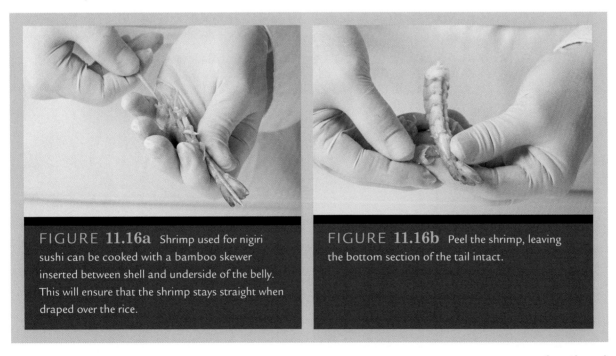

FIGURE **11.16a** Shrimp used for nigiri sushi can be cooked with a bamboo skewer inserted between shell and underside of the belly. This will ensure that the shrimp stays straight when draped over the rice.

FIGURE **11.16b** Peel the shrimp, leaving the bottom section of the tail intact.

(continues)

(continued)

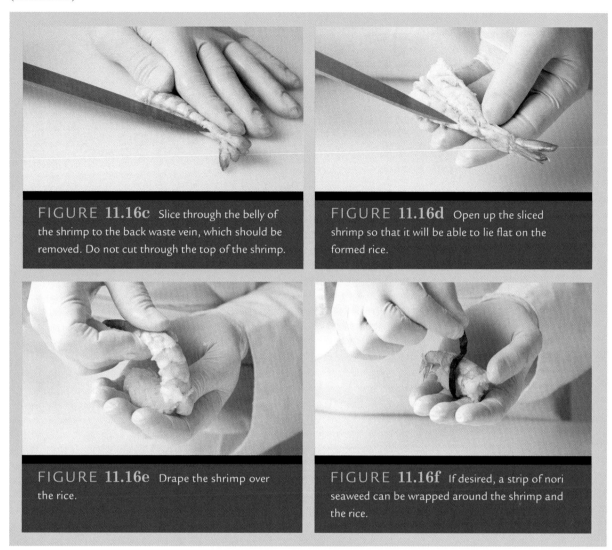

FIGURE **11.16c** Slice through the belly of the shrimp to the back waste vein, which should be removed. Do not cut through the top of the shrimp.

FIGURE **11.16d** Open up the sliced shrimp so that it will be able to lie flat on the formed rice.

FIGURE **11.16e** Drape the shrimp over the rice.

FIGURE **11.16f** If desired, a strip of nori seaweed can be wrapped around the shrimp and the rice.

Maki sushi

FIGURE **11.17a** Cover the bamboo mat with plastic wrap. Place the nori shiny side down on the mat and cover with a thin layer of rice.

FIGURE **11.17b** A variety of fillings can be placed in the center of the rice.

(continues)

(continued)

FIGURE 11.17c Using the bamboo mat, roll the nori around the filling into a tight roll; it can also be shaped into squares.

FIGURE 11.17d Be sure to press down to seal the filling in the nori sheet.

FIGURE 11.17e Cut the ends of the rolls off. The first cut to make rolls is cutting the roll in half.

FIGURE 11.17f Cut it into six or eight pieces depending on the desired size of the sushi.

Maki sushi is one of the most recognized forms. Rice is first pressed onto a sheet of crisp nori seaweed and topped with a wide variety of ingredients, after which it is rolled up with the aid of a bamboo mat and cut into uniform shapes with a sharp knife (see Figs 11.17a through 11.17f). Different sizes and variations have their own names and individual ingredients.

Chirashi sushi

Meaning "scattered" in Japanese, *chirashi* is rice served in a bowl and topped with cooked and raw ingredients. Although referred to as scattered, it is a beautifully arranged dish enhanced by many regional variations.

Temaki

Temaki are made by placing a finger or two of rice and filling ingredients onto a rectangle of nori seaweed, then rolled up into a small cone shape.

PRESENTATION

FIGURE 11.18 Present sushi on rectangular trays accompanied by soy sauce, wasabi, and pickled ginger. Chopsticks should always be raised up off the table.

Japanese-style sushi is a harmony of color, texture, and taste. Simplicity of preparation starts with good knife cuts and proper combinations of ingredients and cooking methods. Plates and platters for sushi are typically square or rectangular and should highlight, not detract from the food (see Fig 11.18). Sushi is always lined up like soldiers on the plate or platter or, in some restaurants, placed directly on the wooden countertop.

CEVICHE

Thought to have originated in Peru, ceviche is a dish or technique used for marinating and denaturing seafood. Its name is possibly derived from the Spanish word *escabeche* meaning "to marinate or pickle in vinegar." The process involves saturating very fresh fish or delicate shellfish with lime, lemon, or other citrus juice. In addition to fresh juices, herbs, onions, or shallots, diced tomatoes and other garnishes can be used. Typically the seafood is sliced very thin, enabling it to denature or "cook" while it marinates. Preparation time depends on the texture and consistency of the product; too long and it will become tough and rubbery. Seafood should be impeccably fresh and held under refrigeration while marinating. Some methods call for the seafood to be quickly blanched before marinating. Other spellings include cebiche, seviche, or cevice, depending on the country.

NUTRITION AND UNDERSTANDING COOKING METHODS AND INGREDIENTS

NUTRITION

For centuries, civilizations labored diligently to supply themselves with ample amounts of wild and domesticated animals and complementary plant proteins. Hard work kept them physically fit and their diet of carbohydrates and fat kept their metabolism in balance. Land animals were leaner; portion sizes were smaller. People grew their own vegetables, ensuring a plentiful supply of this source of nutrients. Processing involved salting, drying, and smoking food. As time went by, we became a sedentary society, resorting too often to fast food, sparking a huge increase in obesity among the general population. The simple diet that nourished our ancestors was no longer the norm; unhealthful food had taken over.

Speculation abounds concerning what percentage of our daily calories should come from protein, carbohydrates, and fat. Diet fads come and go, with all of them proclaiming to have the magic answer. So-called diet experts make exaggerated claims that are confusing and unrealistic.

Good eating habits begin early in life and should include a variety of foods to achieve a proper balance. Many of us were never taught the basics. Studies have shown that by the time children are in second grade, their eating habits are set for life.

Whereas early on they learn that Columbus discovered America and the complexity of the solar system, invariably they are ignorant as to the definition of a calorie and are unable to name the major food groups. Even after graduating college, Americans' knowledge of nutrition is rudimentary, perhaps obtained from TV commercials. We have not made the effort to educate ourselves on the meaning of an age-old axiom, "We are what we eat."

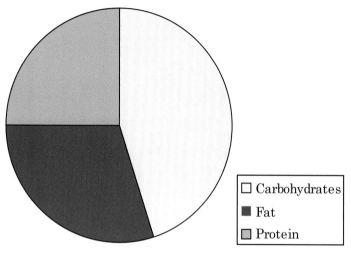

Suggested dietary intakes

- Carbohydrate: 45–65%*
- Fat: 20–35%*
  - Saturated < 10%
  - Polyunsaturated > 10%
  - Monounsaturated > 10%
- Protein: 15–20%*
  - Percent of total calories

CALORIES

Calories are a measurement of energy and, based on gender and activity level, about 2,000 should be consumed each day, a figure that fluctuates according to an individual's energy needs. Expending these daily calories in the right food categories will ensure a nutritious, well-balanced diet.

All foods can be broken down into protein, carbohydrates, and fat; everyone metabolizes each of these differently. The guidelines range from 15 to 20 percent protein, up to 30 percent fat, and 45 to 65 percent carbohydrates. Understanding several simple concepts about each can ensure healthy eating choices.

PROTEIN

Proteins are nutrients essential for the growth and maintenance of our bodies; eight are essential for a healthy life. All proteins are divided into two categories: complete and incomplete proteins. Complete proteins supply all essential amino acids; incomplete proteins lack one or more essential amino acids. The best proteins are lean meats and fish, as well as complementary vegetable proteins such as beans and rice. Consuming a variety of vegetables including grains, cereals, nuts, seeds, and legumes eliminates saturated fats while increasing fiber. Fish high in omega-3 fatty acids is a good source of both protein and "good" fats.

CARBOHYDRATES

Carbohydrates are the most misunderstood category of food. Wheat, rice, and corn account for the majority of the calories consumed worldwide. At least half of our calories should come from carbohydrates, but in recent years, in an attempt to lose weight, many people have pursued a diet low in carbohydrates and high in protein. Carbohydrates fall into two types, simple and complex, both of which are necessary to work efficiently. Simple carbohydrates are found in fruits, juices, dairy products, and refined sugars that are easily absorbed by the body. Because fruits contain nutrients other than simple carbohydrates, they hold a nutritional advantage over other simple carbohydrates, like refined sugars, that provide few nutrients and are considered empty calorie foods. Complex carbohydrates are the most beneficial because they are usually good sources of other important components of a healthy diet such as vitamins and minerals. Complex carbohydrates provide a sense of fullness, and can take up to four hours to digest.

FATS

Fats are essential to a healthy diet and should comprise up to 30 percent of your daily caloric intake. Fats can be found in both plants and animals and come in many forms. As a general rule, fats that are solid at room temperature are not as healthful as those that are liquid at room temperature. Monounsaturated fats (olive oil) and polyunsaturated fats (safflower oil) are two of the best choices. Saturated fats, which should be less than 10 percent of one's fat intake, come from nuts, tropical oils, and animal sources. Fish, low in saturated fat and high in omega-3 fatty acids, are an essential nutrient, and should be included in a healthy diet. Vitamins, minerals, and amino acids are also essential elements and must be obtained from the food we eat; they cannot be manufactured by the body.

OMEGA-3 FATTY ACIDS

Fish and shellfish obtain omega-3 fatty acids from the algae and phytoplankton that they consume. Unfortunately the human body does not have the ability to produce these essential nutrients so we must acquire them from the food we eat. Salmon, herring, sardines, and trout are oily fish containing large amounts of omega-3 fatty acids, which are thought to reduce inflammation and prevent heart disease and arthritis, and inhibit our body's production of specific hormones, which in some people contribute to a wide range of diseases. The American Heart Association (AHA) recommends eating at least two servings of oily fish a week. Other foods that contain large amounts of polyunsaturated fatty acids are dark green leafy vegetables such as spinach, broccoli, and certain nuts and oils including walnuts and canola oil. They have been shown to be quite effective in reducing the risk of heart disease by lowering the amount of cholesterol manufactured in the liver and reducing the likelihood of blood clot formation among deposits of arterial plaque. Omega-3s may also slow or prevent tumor growth, stimulate the immune system, and lower blood pressure.

MERCURY AND SEAFOOD

Mercury is a rare element found in natural deposits worldwide. In its soluble forms, it is toxic and exposure can lead to mercury poisoning presenting in tremors and central nervous system problems. Environmental exposure comes when it is released from coal and gas-fired power plants, industrial manufacturing, and medical and dental applications. It is also used in the manufacturing of electronics, chemicals, and thermometers. Released mercury can accumulate in oceans and streams where it is turned into toxic methylmercury by bacteria in the water. Fish and shellfish absorb and store different amounts of the chemical through absorption, feeding, and age, leading to varying toxic levels. The Food and Drug Administration (FDA) have set a limit for human consumption of methylmercury at 1 ppm.

Mercury levels build up in the bloodstream over time and natural elimination is very slow. Because it can be poisonous to a fetus or young child, the FDA and the Environmental Protection Agency (EPA) recommend that certain species of seafood be avoided by women who are of childbearing age, who are pregnant, or who are nursing. Older fish and shellfish, or fish high on the food chain, typically contain higher amounts because of their prolonged exposure.

The U.S. government recommends seafood as a healthful component of a balanced diet; it is high in protein and nutrients, low in fat. The benefits far outweigh the possible harmful effect of mercury. Recommendations for pregnant women include consuming no more than 12 ounces per week of cooked fish low in methylmercury.

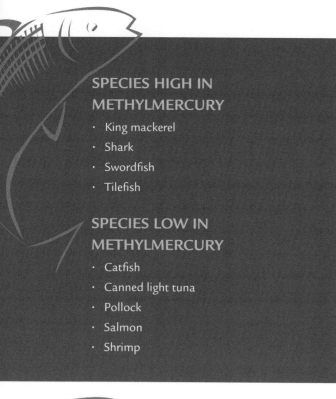

SPECIES HIGH IN METHYLMERCURY
- King mackerel
- Shark
- Swordfish
- Tilefish

SPECIES LOW IN METHYLMERCURY
- Catfish
- Canned light tuna
- Pollock
- Salmon
- Shrimp

A DIET THAT CONTAINS SEAFOOD TWICE A WEEK IS THOUGHT TO

Lower serum cholesterol
Reduce LDL
Lower triglyceride levels
Improve blood platelet function
Be beneficial to many other common diseases

UNDERSTANDING COOKING METHODS AND INGREDIENTS

Seafood can be the easiest or most difficult food to cook. It is high in moisture and protein and varies in fat and texture. It cooks quickly and has a wide range of flavors and textures which, when combined with the correct cooking method and ingredients, results in a great dish. Mastering the cooking methods and understanding ingredients will expand any chef's repertoire far more than individual recipes.

INGREDIENTS

Because it is contrast that brings forth the flavor of the dish, it is important to understand the foods and ingredients that achieve the best results. Flavor is a subtle blend that goes beyond our ability to perceive sweet, sour, salty, and bitter. It is aroma, texture, shape, sight, and sound that make food great. It is the chef's ability to understand and highlight the ingredients of the world's pantry to their utmost potential, individually or as they relate to flavor profiles and cooking methods.

INGREDIENT PAIRINGS

Seafood is delicate and must be paired very carefully with just a few ingredients that will harmonize with and support the item. The objective is to bring out the best flavor and texture, using a minimal amount of accompaniments.

Selection of ingredients should be done very carefully and with a light hand. Combining too many items tends to lose the focus of the food. It is the inexperienced cook who opens the refrigerator and blindly reaches for a multitude of items to create a dish. Fewer, more carefully selected ingredients will work to support the main item and enhance its taste. Remember, it is the piece of seafood that is the focal point, not the sauce or accompaniments.

FLAVOR PRINCIPLE OF TASTE

When chefs speak about a flavor profile or principle, they are referring to the combination of compatible ingredients that create a specific type of flavor. This can be best illustrated with many of the ethnic cuisines of the world. When you taste a chicken stir fry containing garlic, ginger, scallion, and soy sauce, it is clearly Chinese. Mix that same chicken with coconut milk, basil, and curry paste and the flavors resonate more from Thailand. Throughout the Mediterranean region, countries utilize many of the same ingredients but combine them in different ways to create individual and unique flavors. Understanding these ingredients and their pairings is the first step to climbing the staircase of the global kitchen.

FLAVOR COMBINATIONS

Flavor combinations work in many ways. Certain foods have natural affinities to each other; they support one another and make the sum of the two stronger. Others clash and should not be paired. Historically, those items that grew together were served together. A quick look in our summer gardens affirms this. Tomatoes, basil, eggplant, and garlic are natural affinities. Another example is in wild game. When a hunter shoots a deer in the forest, what items can also be found that are edible? Mushrooms, wild berries, leafy greens, and nuts. Isn't it interesting that these are the classical items served with venison? This affinity of ingredients can be a handy tool used to develop great tasting combinations that work extremely well on the plate.

Natural affinities

Ricotta	Honey
Dates	Chorizo
Parmigiano Reggiano	Fruit
Lamb	Rosemary
Goat cheese	Beets
Fish	Citrus
Porcini mushrooms	Rice
Foie gras	Fruit
Apples	Cinnamon

Seafood affinities

Fatty fish

Artichoke, beets, capers, caviar, celery, citrus, fresh herbs, garlic, lemongrass, mayonnaise, nori seaweed, pickled ginger, olive oil, onions, radish, red sweet peppers, saffron, scotch bonnet peppers, sesame, shallots, shiitake and porcini mushrooms, sorrel, soy sauce, spinach, sushi rice, sweet and sour condiments, tamarind, tomatoes, wasabi.

Lean fish

Asparagus, basil, brandy, bread crumbs, butter, celery, chervil, chives, cilantro, citrus, cognac, cream, cucumbers, herbs, leeks, mayonnaise, mushrooms, olive oil, potatoes, saffron, shellfish, tomatoes, vermouth, zucchini.

Pickled fish

Apples, bacon, fresh herbs, mustard, potatoes, pumpernickel bread, sour cream, sweet pickles, toast.

Raw fish

Avocados, carrots, citrus, cucumbers, mayonnaise, nori seaweed, peppers, pickled ginger, pickled vegetables, rice, soy sauce, tomatoes.

Smoked fish

Black pepper, capers, caviar, celery root, cornichons, crème fraîche, dill, hard-boiled eggs, horseradish, lemon, lime, mayonnaise, potatoes, radishes, sour cream, toast.

Shrimp

Bread crumbs, butter, chilies, citrus, coconut, cognac, cream, cucumbers, fermented Asian products, fresh herbs, garlic, ginger, horseradish, lemongrass, miso, mushrooms, olive oil, peppers, Pernod, scallions, sherry, soy sauce, tomatoes, tropical fruits, vermouth, white wine.

Squid and octopus

Citrus, coconut, fresh herbs, garlic, ginger, mayonnaise, peppers, rice, scallions, squid ink, tomatoes, vinaigrette, wine.

Conch

Bread crumbs, celery, chervil, cilantro, citrus, coconut, garlic, mint, parsley, peppers including scotch bonnets, onions, plantains, scallions, tomatoes, vinegar.

Lobster

Brandy, butter, chilies, citrus, coconut, cognac, cream, fennel, fermented Asian products, garlic, ginger, herbs, mayonnaise, mushrooms, mustard, onions, peppers, Pernod, sherry, tomatoes, wine.

Mussels

Bacon, butter, citrus, cream, curry, fennel, fresh herbs, garlic, ginger, leeks, mayonnaise, mushrooms, olive oil, pasta, peppers, Pernod, saffron, scallions, tomatoes, vinegar, white beans, white wine.

Oysters

Bacon, bread crumbs, citrus, cucumbers, curry, cream, fresh herbs, garlic, ginger, horseradish, leeks, peppers, scallions, spinach.

Crabs

Bread crumbs (soft shell), butter, celery, citrus, cream, curry, fresh herbs, garlic, ginger, leeks, lemongrass, mayonnaise, onions, peppers, scallions.

Scallops

Asparagus, avocados, bacon, butter, cabbage, capers, caviar, citrus, cognac, cream, cucumbers, curry, fresh herbs, garlic, ginger, lemongrass, mayonnaise, mushrooms, mustard, olive oil, peppers, Pernod, saffron, scallions, soy products, tomatoes, vinegar, wine.

Sea urchin

Butter, citrus, cream, fresh herbs, nori seaweed, pickled ginger, radish, rice.

SEASONALITY

Freshness and seasonality are a prerequisite for a great dish; in today's global world everything is always in-season somewhere. The concept of seasonality is to use what is in season locally as much as possible, when it is available. Avoid those things that are sourced from long distances or force-ripened.

Change your eating habits to reflect the seasons. In the summer, eat more fresh vegetables and fruits, consume a heavier breakfast and lunch and a lighter dinner. On those long winter nights, gravitate more toward heartier vegetables, beans, and stews. Eating should be less homogenous and more in tune with the world around us.

SEASONAL AND LOCAL SEAFOOD

In the late 1980s, famed chef and restaurateur Gilbert Le Coze left his Brittany home with his sister, Maguy, to open the seafood restaurant Le Bernardin in New York City. The vision was clear; although unusual at the time: develop a menu of all seafood inspired by the foods they were accustomed to eating at their family hotel and their Paris restaurant. In the early months of menu planning, Chef Le Coze made countless trips to the Fulton Fish Market. There he found many of the fish he was familiar with, Dover Sole, Turbot, and Rouget, but there was a small problem, the price. With any restaurant, and especially one featuring all seafood, controlling food cost is critical. Combining this with the high New York City labor cost, the dream of this unique restaurant was in question. After the initial shock, Chef Le Coze had what may have been an epiphany; he realized that some of the freshest and best quality fish were also the most inexpensive. He was intrigued by all the activity in one part of the market and followed the fishy trail to the vendors selling to the Chinese fishmongers who had shops just a few blocks north. Here he found beautiful gleaming fresh fish from the local waters of Brooklyn, New Jersey, and New England. Blackfish, skate wing, white bait, and squid were being sold for incredibly low prices compared to the other varieties. With his cooler full, he taxied back uptown and began subjecting these underutilized beauties to the culinary magic that had earned him his pair of Michelin stars in Paris. It wasn't long afterward that Le Bernardin opened to rave reviews from some of the toughest restaurant critics in the world, New Yorkers. As the years passed, many other high-end restaurants began serving large quantities of fried whole minnows (white bait) and poached skate in brown butter as they began to understand the importance and benefit of using pristine local product.

Seafood is now a global market. Most of the small mom-and-pop operations have been taken over by large grocery store chains that do little or no fabrication and are unable to articulate where or how the fish was caught. Many times customers have to ring the bell and it's the meat butcher who steps in to dispense the seafood. Product has gone from being shipped on ice to being packaged in white plastic containers or frozen in bags and boxes. Very little information is available on the retail level as to when, where, or how the fish was caught. Is it farm-raised, endangered, or flown in from Chile or New Zealand?

There is nothing wrong with frozen seafood when you live in the center of a land-locked state, and many would argue that the quality is better. Regardless of fresh or frozen, all seafood has a season. Fish must spawn to reproduce and taking those fish during that time, or before they are old enough, does not make sense. Understanding what fish are local to your area will ensure freshness and support the local economy. Making smart choices is the best approach with such a perishable and delicate product.

When deciding on ingredient combinations, choose flavors and textures that contrast with one another. A great example of this is a hot dog. Nearly everyone has eaten them slathered with mustard, ketchup, or relish, but why do these flavors work? A hot dog consists of meat, fat, salt, spices, smoke, and moisture. When combined with the sweet or sour flavors in the mustard or relish, the base ingredients taste better. More specifically, the acidity cuts through the high fat content, contrasting it and making it easier to digest. The sugar in the ketchup builds on the sourness of the vinegar and combines with the salt and smoke for a better meat taste. This simple but effective use of supporting ingredients creates a better flavor experience by contrasting each item to its fullest potential. Great taste is achieved in a delicate balance of all the ingredients and cooking methods to create optimum texture and flavor.

KEEP IT SIMPLE

So often this is the hardest thing to do in the kitchen. Chefs train diligently to develop their culinary skills, sharpening them requires a lot of work and dedication. Once they reach the point of conscious competence, they often create dishes that are well executed and properly prepared, but oftentimes complicated, heavily garnished, and containing many overwhelming ingredients and colors. Once chefs realize that more is not better, they can step back and consider what can be removed from the plate, not what can be added. This approach will be ultimately successful because all energy and thought will be focused on the taste and appearance of the main item, enhanced by proper accompaniments. It will also allow for refinements to take the item to the next level.

MISE EN PLACE AND EXECUTION

The best approach to keeping food preparation simple is to understand the cooking methods, execute them properly, and have appropriate *mise en place*. *Mise en place* is a French term referring to work place organization of all ingredients, equipment, and procedures that are required before, during, and after the cooking, allowing for a logical flow without superfluous elements and distractions. Food and flavors are strengthened and highlighted in different ways when they are combined with various methods of cooking, each having its own unique equipment, which should be chosen carefully and used to its potential.

CHOOSING THE CORRECT METHOD OF COOKING

Choosing the proper cooking method begins by understanding the qualities of each species of fish or shellfish. Size, fat content, moisture level, texture, flavor, and color are just a few. Activity level is also important; fish that are the most active have the darkest flesh, the highest amount of fat, and the most intense flavor. Approaching these differences is best done through contrast and balance. Lean species such as flounder are often cooked with the addition of fats, using cooking methods such as sautéing, poaching, deep frying, or pan frying, and are accompanied by richer sauces such as remoulade and Vin blanc. Tuna, which is high in fat, is typically not deep fried, but grilled, and contrasted with simple sauces such as relishes and chutneys. The fresh flavor of these sauces and the charring of the grill support and highlight the tuna to create a balanced dish.

COOKING METHODS

Cooking methods encompass two important elements: type of heat used and the selection of the proper cooking vessel for the ingredients involved. Each method delivers a different flavor and texture and should be chosen to highlight the food's characteristics by applying the best possible cooking application. In this chapter, the cooking methods will be broken down into four fundamental categories.

1. Dry-heat cooking without fats and oils—These methods rely on dry heat without the addition of fats or oils. The food is cooked either by direct application of radiant heat (as with grilling and broiling) or by indirect heat in an oven (roasting and baking), and achieves a highly flavored exterior and a moist interior.

2. Dry-heat cooking with fats and oils—These cooking techniques use fat or oil as the cooking medium. Results vary, depending on the amount of fat used.

3. Moist-heat cooking methods—Items prepared using moist-heat cooking techniques result in products that have a distinctly different

flavor, texture, and appearance from those prepared with the dry-heat method. These techniques typically require the use of naturally tender meats, poultry, or fish, and vegetables and fruits. The proper selection of a flavorful liquid adds an important dimension to many of these preparations. Careful monitoring of cooking temperatures and times and the ability to determine doneness are also critical to a mastery of moist-heat methods.

4. Combination cooking methods—Braises and stews are often thought of as peasant dishes because they frequently call for less tender (and less expensive) main ingredients than other techniques. These dishes have a robust, hearty flavor and are often considered fall and winter meals. However, by replacing traditional ingredients with poultry, fish, or shellfish, braises and stews can be faster to prepare, lighter in flavor and color, and appropriate for contemporary menus.

GRILLING AND BROILING

Grilling and broiling rely on a direct application of high radiant heat to quickly cook naturally tender, portion sizes of meats, poultry, fish, and vegetables.

Grilled foods utilize a heat source from below the foods and should have a smoky, slightly charred flavor resulting from the flaring of the marinades, juices, and fats that are rendered out of the item as it cooks. Broiling is similar to grilling but uses a heat source located above the food rather than below. Frequently, delicate lean white fish are brushed with butter or seasoned oil, put on a heated, oiled sizzler platter, and then placed on the rack below the heat source instead of directly on the rods.

Grills come in many shapes and sizes depending on the volume and application. One with heavy cast iron grill bars is the best choice; it will conduct heat efficiently and will be easy to clean. Many grills have lava rocks or flavor bars below the cooking surface and above the heat source. These act to capture the drippings, avoiding flare-ups and generating flavorful smoke which gives the food its unique "grilled" taste. Most commercial grills are fueled by gas or electric, but there has been a resurgence of wood grills, which are excellent for flavor, but difficult to regulate, and depending on the price of the wood, can be costly to operate. Some gas or electric grills have a wood box that can be filled with wood chips or sawdust to enhance flavor development. Special woods such as mesquite, hickory, or fruitwoods are frequently used to impart a special flavor. Other aromatic items such as herb stems, grapevines, and teas can be soaked in water and placed on the grill to create a flavorful smoke. The temperature of the grill can be controlled by grill zones of varying heat intensity, including a very hot section for quickly searing foods and an area of moderate to low heat for slow cooking or holding foods. Zones may also be allocated for different types of foods, to prevent an undesirable transfer of flavors. Developing a system for placing foods on the grill or broiler, whether by food type or by degree of doneness, helps speed work on a busy line or allows the home cook a more professional approach to cooking.

Tender, portion-sized cuts are important to grilling, and many varieties of seafood can be grilled with excellent results. Firm and fatty fish work best, as do nearly all shellfish. Because appearance of the main item is the first step to proper presentation, fabricate and trim the item in an attractive shape before cooking.

Proper mise en place is important to all cooking methods. Grills and broilers must be well maintained and kept clean to produce good-quality grilled or broiled food. Take the time to prepare the grill by first heating it up to high; thoroughly clean the bars with a sturdy grill brush. Oftentimes it is necessary to lubricate the rods throughout a busy service time to keep items high in moisture from sticking. Metal skewers need to be cleaned and oiled before use, and tongs, spatulas, sizzle pans, brushes, or barbeque "mops" are all part of the station's equipment mise en place. Hand racks for delicate foods or those that might be awkward to turn easily, such as whole fish, should also be cleaned and oiled between uses to prevent the skin of the fish from sticking and tearing.

Quality characteristics

Foods that are properly grilled and broiled have a distinctly smoky flavor, which is enhanced by charring and accompaniments that contrast and elevate the flavors of lean meats and fatty fish. This smoky flavor and aroma should not overpower the food's natural flavor, and the charring should not be so extensive that it gives the food a bitter or carbonized taste. Any marinades or glazes should support, not mask the natural flavor.

Marinades

Because foods that are grilled or broiled are naturally lean, they are often marinated prior to cooking. Marinades generally contain one or more of the following: oil, acid, and aromatics. Oils protect food from intense heat during cooking and allow other flavorful ingredients to remain in contact with the food. Acids, such as citrus juices, vinegar, wine, and yogurt flavor the food and change its texture. In some cases, acids firm or stiffen foods (e.g., the lime juice marinades that "cook" the raw fish in ceviche); in others, they break down connective fibers, making tough cuts of meat more tender. An example is using a red wine marinade over several days when preparing beef bourguignon. Aromatics provide specific flavors.

> **HOW TO GRILL AND BROIL**
>
> 1. Place the seasoned food, presentation side down first, on the preheated grill or broiler rods to mark it and to start the cooking. When the food comes into contact with the heated rods, marks are charred onto the surface of the food. To mark foods with a crosshatch on a grill or broiler, turn it a quarter turn.
> 2. Because many barbecue sauces contain sugar and burn easily, wait until the food is almost completely cooked before applying the sauce. As the food finishes cooking, the sauce glazes and caramelizes lightly without burning.
> 3. Turn the food over and continue cooking to the desired doneness. Because most foods cooked by grilling or broiling are relatively thin and tender, once turned over, they require minimal cooking time. Thicker cuts such as fish steaks can be moved to a cooler area of the grill or broiler to avoid a charred exterior. For banquets, foods can be quickly marked on the rods of a grill or broiler, laid out on racks over sheet pans and finished in the oven, maximizing output of the grill or broiler. For food safety reasons, exercise care in not chilling the food too quickly, and monitor the fish's internal temperature with a sterile thermometer.

Marinating times vary according to the food's texture. Tender or delicate foods such as fish or poultry breast require less time. Highly acidic marinades may take only 15 to 20 minutes, whereas others are best when left for several hours.

Some marinades are cooked before use; others are not. Sometimes the marinade is used to flavor an accompanying sauce or may itself become a dipping sauce. Marinades that have been in contact with raw foods can be used in these ways provided they are boiled first to kill any harmful pathogens.

To use a liquid marinade, add it to the item and coat it evenly. To marinate is to coat, not submerge; once the item is submerged it will begin to absorb the liquid and change its flavor and texture. Brining is the submerging technique using water, salt, sweetener, and spices.

Characteristics of grilled or broiled items
- Tender
- Well marbled, or for fish, high fat content
- Individual portions

SAUTÉING

Sautéing is a technique that cooks food rapidly in a small amount of fat over relatively high heat. The term *sauté* comes from the French verb *sauter*, meaning "to jump," and refers to the way foods sizzle and jump in a hot pan. Certain menu items listed as seared or pan seared are also essentially sautés and often are menu marketing terms, or may indicate that the food is cooked extremely rare. Searing is also done to large cuts of meat as a first step prior to roasting, braising, or stewing. Sautéed dishes typically include a sauce made with the drippings, or fond, left in the pan.

Foods that are sautéed are naturally tender and include all types of seafood. Many delicate vegetables such as zucchini and snow peas are appropriate for sautéing.

Choose the cooking fat according to the nutritional value, flavor, application, and food cost. Oils such as olive, corn, canola, soy, and clarified butter are the most common. Rendered fats, such as lard, bacon, and goose fat are not desirable due to lack of availability and health concerns. The base for a pan sauce in sautéing may vary to suit the flavor of the main item. Frequently it is "built" using the fond as the initial element. Aromatics consisting of shallots, leeks, onions, garlic, or tomatoes are added to the fond and any remaining fat to establish the base of the sauce. Deglazing with wines, stock, vegetable juices, or liquor adds to the viscosity, especially after they get reduced, to intensify flavor and thicken the sauce. Many premade sauces, such as velouté, béchamel, or tomato can be added as well. Classically, many of the fine sauces were finished with butter and strained giving them a smooth, rich, and delicious flavor.

Quality characteristics
Sautéed foods should have a flavorful and caramelized exterior, resulting from proper browning, which serves to intensify the food's flavor. Weak flavor and color indicate that the food was sautéed at too low a temperature or that the pan was too crowded. Proper color depends on the type of food. Lean white fish should be pale gold when

HOW TO SAUTÉ

1. Because sauté is a quick-cooking technique, proper mise en place is critical to a successful outcome. Once all the ingredients and equipment have been gathered, fabricated, and measured, the cooking can begin.

2. To begin the flavor development, protein items should always be seasoned with salt and pepper prior to cooking. If seasoning is done only after the item is cooked, taste is compromised. Kosher salt is preferred in the professional kitchen; its large flat-sided crystal shape adheres to food better than regular table salt. Dredging the item in flour prior to placing it in the pan will prevent it from sticking and aids in caramelization and fond development.

3. Select a pan that is very heavy and large enough so the main item covers the bottom of the pan without overlapping. Avoid non-stick pans unless they are used specifically for items that will stick or if a reduced fat sauté is required. Heating the pan before adding oil is referred to as "conditioning the pan." Add enough fat to lightly coat the bottom of the pan and wait until it begins to shimmer.

4. Immediately add the food to the pan, placing the presentation side down. Let the food cook undisturbed for several minutes or until the bottom is golden brown and the edges begin to crisp. Heat temperature can vary depending on the type of food being cooked.

5. Turn the sautéed foods once they have achieved a consistent golden color and cook until a proper internal temperature is reached. In some cases food may be finished in the oven either in the pan or on a sizzle platter.

6. Once the food has been removed from the pan, begin to add aromatic ingredients to make the sauce. These often include garlic, onions, shallots, leeks, or tomatoes, items that are sweated or lightly browned depending on the finished sauce. Next deglaze the pan with wine, stock, cream, or other liquid to form the base of the sauce. Wines used for deglazing are typically reduced until almost dry prior to adding other ingredients. Thickening is accomplished by reduction, or with a slurry of starch and liquid, or with a premade thickened sauce.

7. Once the sauce is properly reduced, it can be finished by adding appropriately cut vegetables or herbs as garnish. It may be strained through a fine-meshed strainer for a very smooth texture. Add finishing ingredients and simmer until properly cooked and heated. Adjust seasoning with salt, pepper, fresh herbs, juices, or a small amount of butter to add flavor, body, and a glossy appearance. At this point, the main item can be returned to the pan for reheating.

sautéed as skinless fillets, or golden brown and crispy if served with the skin on. Firm and fatty fish will take on various darker colors depending on the species and meat characteristics.

Only naturally tender foods should be sautéed, and after sautéing they should remain tender and moist. Excessive dryness is a sign that the food was overcooked, or that it was cooked too far in advance.

PAN FRYING

Pan-fried foods have a crisp textured crust and a moist flavorful interior, producing a dish of contrast and flavor that is served with a separately made sauce.

Pan-fried food is almost always coated in breading or dipped in batter. It is fried in enough oil to come halfway or two-thirds up its side, often over less intense heat than that in sautéing. The product is cooked more by the oil's heat than by direct contact with the pan. In pan frying, the hot oil seals the food's coated surface, trapping the natural juices inside. Because no juices are released and a large amount of oil is involved, accompanying sauces are usually made separately.

Pan-fried food is usually portion size or smaller. Fish and shellfish are well suited to this technique, and any fat, skin, or bones should be removed. Certain firm seafood such as shrimp and tuna can be pounded out into malleable thin sheets suitable for rolling or ceviche. Rolled portions are easy to bread and cook up crisp and flavorful. Be certain to season the food before adding a coating.

Proper mise en place for a breading station includes separate containers of flour, beaten eggs, and a breading material, which can be fresh or dried bread crumbs, cornmeal, or crushed crackers. In this set up, the seasoned product first gets dredged in flour followed by the egg wash and finally the breading.

The fat for pan frying must be able to reach high temperatures without breaking down or smoking. Commercial frying oil burns longer and cleaner than animal fats and has a neutral flavor, without the saturated and trans fats. The choice of fat makes a difference in the flavor of the finished dish; keep it clean and take safety precautions when working with heat and oil.

The pan used must be large enough to hold foods in a single layer without touching one another. If the food is crowded, the temperature of the food will drop quickly and a good crust will not form. Pans should be made of heavy-gauge metal and should be able to transmit heat evenly. The sides should be higher than those appropriate for sautés to avoid splashing hot oil out of the pan as foods are added to the oil or turned during cooking. Because the amount of oil needed for a pan-fry item is so great, a sauce is typically made and served separately from the item in order for it to maintain its texture.

FIGURE 12.1 Standard breading station setup includes flour, egg wash, and bread crumbs. It is always a good idea to keep the egg wash on ice. Keep one hand for all dry work and one hand for all wet work; otherwise your fingers will become breaded.

Standard breading

Standard breading is a method of coating foods with flour, egg wash, and bread crumbs to create a crispy crust on fried foods (see Fig 12.1). Following a proper breading procedure will ensure that the finished product is completely enrobed in an expedient and efficient manner.

Flours and similar meals or powders such as cornstarch are used to lightly dredge or dust foods before they are dipped in an

HOW TO PAN FRY

1. Fabricate, trim, and, if necessary, pound or flatten the item to aid in cooking and presentation. Season with salt and pepper.
2. Coat the food in a standard breading or batter of your choice.
3. Heat the pan and add the proper amount of fat. As a rule of thumb, add enough fat to cover one third to one half of the sides of the food; the thinner the food, the less fat required. When a faint haze or slight shimmer is noticeable, the fat is usually hot enough. To test the temperature, dip a corner of the food in the fat. If the fat is at about 350°F/177°C it will bubble around the food and the coating will start to brown in less than a minute.
4. Add the food carefully, dropping it away from you to avoid any splatter and prevent burns. Achieving an evenly browned and crisp food requires that the food be in direct contact with the hot fat. If the foods are crowded, they may not develop good color and texture. If there is not enough fat in the pan, the food may stick to the pan and tear, or the coating may come away. When pan frying large quantities, skim or strain away any loose particles between batches and add more fat as needed.
5. Turn the food once and continue to pan fry until the second side is golden and the food is properly cooked. Some thick steaks or fillets may need to be removed from the fat and placed in an oven to finish cooking. In that event, be sure that they are not covered, as this could create steam and make the food soft and unappealing. Determining doneness is difficult with foods that are thin, but the accepted method is to insert a sterilized instant read thermometer into the thickest part of the food.
6. Once the food is properly cooked, serve it immediately or transfer it to a wire rack placed on a sheet pan and hold briefly in a warming oven for service.

COOKING TEMPERATURES FOR SEAFOOD

Because seafood is high in water and has different protein structure than land animals, the way it reacts to heat and cooking temperatures is unique.

- *Medium rare seafood   120°F/49°C
- Fully cooked seafood   145°F/63°C

* The Food and Drug Administration (FDA) suggest that it is always best to thoroughly cook all seafood to minimize the risk of foodborne illness. If it is to be eaten raw then it should have been previously frozen.

KEEP IT SIMPLE

egg wash. Egg wash is made by blending eggs and milk or cream at a ratio of 2 parts eggs to 1 part milk. Some items such as calamari are dipped into milk, buttermilk, or beer before applying breading, rather than using egg wash.

Bread crumbs may be dry or fresh. Fresh white bread crumbs (called *mie de pain* in French) are prepared by grating or processing a finely textured white bread with the crust removed.

Dry bread crumbs are prepared from stale bread that is processed into a fine texture meal. Asian-style bread crumbs, also known as *panko,* are produced commercially, using a special process that creates a large, pure white, almost crystal-like crumb that maintains its structure and texture throughout the cooking process. Becoming increasingly popular worldwide, panko crumbs are very good for deep-fried shrimp, and can be purchased in Asian grocery stores.

Other ingredients may be used in place of or in addition to bread crumbs. Options include nuts, seeds, shredded coconut, corn flakes, potato flakes, shredded potatoes, grated cheese, and chopped herbs.

Blot the food dry with absorbent toweling and season as desired. Hold it in one hand and dip it in flour. Shake off any excess flour and transfer the food to the container of egg wash. Switch hands, pick up the food and turn it if necessary to coat it on all sides. Transfer it to the container of bread crumbs and lightly press the crumbs evenly around the food. Shake off any excess and transfer to a sheet pan with parchment paper between layers. Cook immediately or refrigerate briefly to avoid soggy breading.

Discard any unused breading because the presence of juices and drippings will contaminate the product and make it unsafe for use with other foods.

Quality characteristics

The object of pan frying is to produce a flavorful exterior with a crisp, brown crust, which acts as a barrier to retain juices and flavor. Because the food itself is not browned, the flavor will be different than if it had been sautéed. The color depends on the coating or batter, but all pan-fried foods should be beautiful golden brown and crispy on the outside and tender in the middle. A pale or inconsistent color indicates that the oil amounts were wrong or incorrect heat levels or pan sizes were used. Sauces used for pan-fried foods are made separately and typically contrast the lean flavors of the protein item or the crispy texture of the coating. Lemons, rémoulade, or mayonnaise-based sauces are popular for pan-fried fish dishes.

DEEP FRYING

Deep-fried foods have many of the same characteristics as pan-fried foods, including a crisp, browned exterior and a moist, flavorful interior. Deep-fried foods, however, are cooked in enough fat or oil to completely submerge them. The food is almost always coated with a standard breading or batter such as a tempura or beer batter or a simple flour coating. The coating acts as a barrier between the fat and the food and also contributes flavor and texture contrast.

Select foods that are naturally tender. Be certain to season the food before adding a coating. Deep frying is also suitable for croquettes made from any farinaceous product. They are typically garnished with salt cod, clams, shrimp, vegetables, and fish.

Professional gas or electric deep fryers with baskets are typically used for deep frying, although it is also feasible to fry foods using a large pot. The sides should be high enough to prevent fat from foaming over or splashing, and wide enough to allow the chef to add and remove foods easily. Use a deep-fat frying thermometer to check the fat's temperature, regardless of whether you use an electric or gas fryer or a pot on a stovetop. Become familiar with the fryer's recovery time; that is, the time needed for the fat to regain the proper temperature after foods are added. The fat will lose temperature for a brief time. The more food added, the more the temperature will drop and the longer it will take to come back to the proper level. Of all the equipment in the kitchen, the deep fryer is the most dangerous and must be used with caution to avoid splatter and fires, especially in the home kitchen.

Quality characteristics

Deep-fried foods should be crispy on the outside and moist and flavorful on the interior. They should taste like the food without any trace of foods previously fried in the fat. Foods served very hot, directly from the fryer, have a better taste. If the food tastes burnt or greasy, the fat was not hot enough or too old.

STIR FRYING

Stir frying is a quick-cooking Asian technique utilizing high heat and naturally tender foods. Originating thousands of years ago in China, it evolved from the limited availability of fuel and high cost of suitable protein. The main cooking vessel is the *wok,* which means "pot" in Cantonese. The wok is a very versatile bowl-shaped piece of equipment that can also be used for steaming, frying, braising, stewing, and smoking. Unlike the sauté pan, the wok is very thin, does not conduct heat as well, and must be stirred using two long tools designed to move the ingredients efficiently throughout the varying temperatures of the wok. Classic wok-cooked dishes combine centuries of culinary learning and encompass a true balance of flavors, textures, consistencies, and a sound understanding and appreciation of the earth's bounty and one's responsibility to it.

HOW TO DEEP FRY

1. Heat the cooking fat to the proper temperature which is generally 325° to 360°F/163° to 182°C. The cooking fat must reach and maintain a nearly steady temperature throughout frying time to prepare crisp, flavorful, and nongreasy fried foods. Proper maintenance of oil will help extend its life. Old fats and oils have a darker color and more pronounced aroma than fresh oil. They may also smoke at a lower temperature and foam when foods are added. Be sure to strain or filter the oil properly, as needed, based on use.

2. Two methods are used to deep-fry foods. The choice depends on the food, the coating, and the intended results. The swimming method is generally used for battered food. As soon as the food is coated with batter, it is carefully lowered into the hot oil using tongs. At first the food will fall to the bottom of the fryer; as it cooks it swims back to the surface. For even browning, it may be necessary to turn foods once they reach the surface for them to brown evenly. The basket method is generally used for breaded items. Place the breaded food in a frying basket and then lower both the food and the basket into the hot fat. Once the food is cooked, use the basket to lift out the food. Determining doneness is best accomplished with a sterilized instant-read thermometer into the thickest part of the food. Deep-fried food should remain uncovered and served immediately.

Because the actual cooking takes place so quickly, it is extremely important to have all the ingredients prepared and assembled ahead of time. Thorough mise en place is critical to a successful outcome.

Foods that are stir fried are naturally tender and include all types of seafood as well as delicate vegetables such as zucchini and snow peas. Foods are customarily cut into small pieces, usually strips, dice, or shreds, and cooked rapidly in a little oil. They are added to the wok in sequence; those requiring the longest cooking times are added first, those that cook quickly only at the last moment. Most Asian-style stir frys begin with a mixture of garlic, ginger, and scallions, or GGS. The oil used should be suitable for high-heat cooking; typically peanut or soybean oil is used.

Many stir fry dishes are very nutritious because they are largely plant-based and can be cooked using very little fat. Sauces are normally thickened with slurry of cornstarch and liquid and should be of sufficient amount and thickness to coat the food without appearing to be soupy. Protein items such as shellfish can be marinated and "velveted" prior to cooking for flavor and texture. Velveting requires tossing the specific item in a mixture of egg whites and cornstarch. These ingredients act to season and coat the product and give it a soft delicate texture to which the sauce can adhere. This technique can also be accomplished by plunging the food into oil that is between 260° to 300°F/127° to 149°C, but it adds unnecessary fat that can be very unpleasant.

HOW TO STIR FRY

1. Assemble all the equipment and ingredients needed to cook the dish, including a serving platter.
2. Cut the vegetable and protein items in consistent shapes and blanch any hard vegetable that will not cook in a short amount of time. If desired, velvet the main item and let it sit about 20 minutes.
3. Heat the wok over very high heat; a ring can be used to hold the wok in place over the flames.
4. Add a small amount of oil to the hot wok and immediately add the GGS, working it rapidly around the sides with the tools.
5. Add the protein items and stir fry until about halfway cooked.
6. Add the other ingredients in order of hardness from hardest to softest. Continue to stir fry until the vegetables are tender and not overcooked.
7. Add the selected sauces and thicken with slurry.
8. Finish with sesame seeds or oil and a pinch of sugar to round out the flavor.
9. Accompany the stir fry with rice or noodles.

Quality characteristics

The objective of stir frying is to produce a flavorful combination of ingredients that are bound together with a lightly thickened sauce and cooked very quickly over high heat. Quality characteristics include bite-sized pieces of seafood that are highly flavorful, tender, and free of skin or bones. Vegetables should be served in abundance and should retain their integrity and color.

POACHING

Because of the delicate nature of seafood, poaching is an excellent technique with lots of flavor and nutritional potential. It is broken down into two separate methods, shallow and deep poach; each has different outcomes.

Shallow poaching

Shallow poaching, like sautéing and grilling, is an a la minute technique. Foods are cooked partially submerged in a combination of steam and simmering liquids such as wine or lemon juice. Aromatics, such as shallots and herbs, are added for more flavor. The pan is loosely covered with parchment paper to capture some of the steam released by the liquid during cooking; there is a significant amount of flavor transfer from the food to the cooking liquid. For maximum flavor, the cooking liquid (cuisson) is usually reduced and used as the base for a sauce. The wine and lemon juice give the sauce a bright, balanced flavor. Butter can be easily emulsified in the sauce, making beurre blanc, or a premade velouté sauce can be added to the reduced cooking liquid.

Naturally tender lean fish and shellfish are among the most common options for this cooking method. Trim the product as appropriate, and remove bones or skin from the fish fillets. Fish fillets may be rolled or folded around a stuffing to form paupiettes, with the skin side on the inside.

The liquid contributes flavor to the food as well as to the sauce prepared from the cooking liquid. Choose rich broths or stocks and add wine, vinegar, or citrus juices to expand the flavor profiles of the dish. Cut aromatics fine or mince them. Other ingredients to be served along with the sauce as a garnish should be cut neatly into attractive cuts to highlight the skill level of the dish and for consistency of cooking.

> **HOW TO SHALLOW POACH**
>
> 1. Lightly butter a shallow pan and add aromatics to enhance the flavor of the cooking liquid and finished sauce. If they can cook completely in the time required, they can be added raw. Otherwise, cook them separately beforehand or sweat them in butter.
>
> 2. Place the seasoned food on top of the aromatics, and then pour the liquid around the item. Except for large quantities, preheating the liquid is not necessary. Be careful not to have it at full boil.
>
> 3. Bring the liquid to a gentle simmer over low heat. Loosely cover the pan with parchment paper and finish cooking in a moderate oven or on the stove top. Cooking temperatures differ with ingredients, but most seafood should be cooked between 140° and 180°F/60° and 82°C.
>
> 3. Remove the food when it is several degrees below the desired temperature to allow for carryover cooking. Cover the food loosely and keep it warm.
>
> 4. The sauce is typically made by reducing the cooking liquid and whipping in softened butter, or by adding a thick crème fraîche or a premade velouté sauce.

The sauce may be a beurre blanc or vin blanc, or simply the reduced cooking liquids served as a broth. Shallow poaching is done in a sauté pan or other shallow cooking vessel, such as a sautoir or rondeau. Select the pan or baking dish carefully. Improper space around the food will result in overcooking or undercooking and affect the amount of liquid available for the sauce. Buttered parchment paper is generally used to loosely cover the pan as the food cooks. It traps enough of the steam to cook, but not so much that the cooking speed changes.

QUALITY CHARACTERISTICS: When well prepared, shallow-poached dishes reflect the flavor of both the food and the cooking liquid. The flavor is a bright balance of acidic and aromatic ingredients. The sauce adds a rich complementary flavor, and should appear moist, opaque, and pale-to-light in color. The sauce should have delicate velvet-like shiny

> **HOW TO DEEP POACH**
>
> 1. Combine the food with the liquid and bring to the correct cooking temperature. Some foods are started off in cool liquid, whereas others are lowered into liquid that is already at the correct temperature. Poaching liquid should range from 140° to 180°F/60° to 82°C.
> 2. Maintain a consistent cooking temperature throughout the poaching process and avoid using a lid, which tends to increase the pressure and boil the liquid.
> 3. Determining doneness is best accomplished with an instant-read thermometer. Once cooked, poached food should be served immediately with the appropriate sauce.

texture, and fresh herbs must be aromatic and green. Classically, these sauces are strained and any additional garnishes would be added right before plating.

Deep poaching

Deep poaching is a cooking technique that requires the food to be completely submerged in a liquid kept at a constant, moderate temperature. The aim is to produce foods that are moist and extremely tender. The distinguishing factors between the two methods are differences in cooking temperature and appropriate types of food. Deep poaching is perfect for seafood and poultry and is done at low temperatures.

Naturally tender fish items to be poached can be left whole or cut into portion size. Dressed fish should be wrapped in cheesecloth to protect it from breaking apart during cooking.

The liquid used in deep poaching should be well-flavored; for seafood a fish stock, fumet, wine, or a court bouillon is appropriate.

Aromatic ingredients such as herbs and spices, wine, vegetables, vegetable juice, or citrus zest may be added to the cooking liquid to enhance the flavor of the finished dish. Sauces are often served separately. Poached salmon, for instance, is traditionally served with a warm butter emulsion sauce, such as béarnaise or mousseline sauce. See specific recipes for sauce suggestions.

Because the food is to be submerged in liquid, a tall-sided sauté pan or fish poacher is often used, and a good thermometer that will continually monitor the temperature will ensure a perfectly cooked dish.

QUALITY CHARACTERISTICS: When properly cooked, poached foods are nearly colorless, fork tender, and extremely moist. Fish and shellfish should be slightly firm and lose their translucency. Shellfish open and the edges of the flesh curl. Shrimp, crab, and lobster have a bright pink or red color. Flavor, appearance, and texture are all important. In an ideal balance, the aromatics, seasonings, and flavorings either bolster or complement the flavor of the food.

STEWING

Stews are dishes that have a robust, hearty flavor and are wonderful one-pot meals, which can be made from nearly all foods. The size of the cuts will vary according to the style of stew, but typically they are bite-sized cubes. Season foods for stewing before cooking, using salt, pepper, marinades, and dry rubs to give the finished dish a good flavor. Select the cooking liquid according to the food being stewed or the recipe's recommendations. Flavorful shellfish stocks and sauces or vegetable juices can be used.

Stews often include vegetables, both as an aromatic component and as an integral component of the dish. Rinse, peel, and cut vegetables into uniform shapes so that they will cook properly. Keep the vegetables separated so that they can be added to the stew in the proper sequence. Choose a heavy-gauge pot with a lid for slow, even cooking.

Quality characteristics

When done, a stew is extremely tender, almost to the point where it can be cut with a fork, but not where it falls into shreds. Every component of the stew should be fully cooked and tender. Discard any sachet d'epices or bouquet garni. It should have a rich flavor and soft, almost melting texture. The natural juices of the ingredients, along with the cooking liquid, become concentrated and provide both good flavor and a full-bodied sauce. The major components in a stew retain their natural shape, although a certain amount of volume may be lost during cooking.

HOW TO STEW

Seafood items should not be browned in the initial steps, but can be seasoned and marinated for flavor.

1. Heat the pot and cook the aromatics and vegetables in oil according to the specific recipe. Add wine or other liquid and reduce. Add flour to thicken the stew and then add enough liquid to cover the vegetables.
2. Cook the vegetables until tender.
3. Add the seafood and more liquid to cover.
4. Cook gently until the seafood is cooked. Finish with fresh herbs.

13

RECIPES

Ceviche **275**

Salt Cod Fritters **276**

Conch Fritters **277**

Crab Cakes with Creole Honey-Mustard Sauce **279**

Creole Honey-Mustard Sauce **281**

Garlic Shrimp **282**

Clam Sauce **283**

Egg Pasta **284**

Shrimp Tempura **285**

Tempura Dipping Sauce **286**

Sushi Rice **287**

Japanese Hand Vinegar **288**

Wasabi **288**

Nigiri Sushi **289**

Japanese Omelet for Nigiri Sushi **290**

Nori-Roll Sushi **291**

Miso Soup **293**

Dashi **294**

Shrimp Bisque **295**

New England Clam Chowder **296**

Mussels Marinière **297**

Bouillabaisse **299**

Rouille **301**

Salade Niçoise **303**

Trout with Sautèed Mushroom **305**

Dover Sole Meunière **307**

Shrimp with Tomatoes, Feta, and Oregano **309**

Paella **311**

Base Recipe for Shallow Poached Fish **315**

Fillet of Flounder with White Wine Sauce **317**

Grilled Salmon with Ginger Glaze **319**

Pan-Fried Cod **321**

Fish and Chips **323**

French Fried Potatoes **325**

Lobster Thermidor **326**

Cold-Smoked Salmon **327**

Gravlox **328**

Salmon Rillette **329**

Fish Stock **330**

Mousseline Forcemeat **331**

CEVICHE

YIELD: 10 APPETIZER PORTIONS

Ingredient	US	Metric
Scallops, sliced very thin	1 lb	454 g
Flounder, sliced very thin	1 lb	454 g
Wild shrimp, sliced very thin	8 oz	227 g
Plum tomatoes, peeled, seeded, brunoise	2	2
Cucumbers, peeled, seeded, brunoise	1/2	1/2
Minced chives	2 tsp	10 mL
Chopped cilantro	1 Tbsp	15 mL
Chopped oregano	1 tsp	5 mL
Minced garlic	2 tsp	10 mL
Poblano pepper, brunoise	1/2	1/2
Yellow bell pepper, brunoise	1/2	1/2
Jalapeño, minced	1/4–1/2	1/4–1/2
Olive oil	1/4 cup	60 mL
Ground coriander seed	1/2 tsp	2.5 mL
Ground cumin seed	1/2 tsp	2.5 mL
Sugar	1/4 tsp	1.25 mL
Hot sauce	5 drops	5 drops
Lime juice, or as needed	8 oz	227 g
Kosher salt	as needed	as needed
Crusty bread, thinly sliced	20 slices	20 slices

METHOD

1. Combine the scallops, flounder, shrimp, tomato, cucumber, herbs, garlic, peppers, oil, spices, sugar, and hot sauce in a large mixing bowl.
2. Add enough lime juice to cover the seafood. Season with salt as needed.
3. Refrigerate and marinate at least 8 hours, or until the fish is de-natured and bright white but not tough.
4. Arrange attractively on an appetizer plate or in a martini glass, and garnish with sliced bread.

NOTE: Because this preparation is denatured by acidity and not cooked with heat, it is extremely important to use live shellfish and very fresh fish.

SALT COD FRITTERS

YIELD: 10 PORTIONS

Salted cod, boneless	2 lb	907 g
Sofrito		
Vegetable oil	1/4 cup	60 mL
Garlic cloves, chopped	8	8
Onion, small dice	4 oz	113 g
Red bell peppers, small dice	2	2
Green bell peppers, small dice	2	2
Fritter Batter		
All-purpose flour	12 oz	340 g
Baking powder	4 tsp	20 mL
Water	1 3/4 cup	420 mL
Chives, chopped	1/2 bunch	1/2 bunch
Green onions, green parts only, sliced on bias	1/2 bunch	1/2 bunch
Ground black pepper	1 Tbsp	15 mL
Frying oil	as needed	as needed

METHOD

1. Cover the salt cod with water and place in the refrigerator. Soak the cod for up to 24 hours, or until the salt has leached out. Drain the salt cod and pat dry. Flake the cod finely and reserve.
2. For the sofrito: Heat the vegetable oil in a large skillet over low heat and add the garlic, onions, and red and green peppers. Sauté until soft, about 15 minutes. Allow to cool to room temperature.
3. For the batter: Combine the flour and baking powder in a medium-size bowl and add the water. Mix thoroughly with wire whisk until the batter is smooth, without any lumps. Fold in the chives, green onion, black pepper, flaked cod, and sofrito.
4. Heat the oil to 350°F/177°C. To form the fritters, drop the mixture 1 tbsp/15 mL at a time into the hot oil and fry until golden brown, 2 to 3 minutes. Drain on paper towels and serve hot.

CONCH FRITTERS

YIELD: APPROXIMATELY SIXTY 2 oz/57 g FRITTERS

Conch, ground through an 1/8-inch die	4 lb	1.81 kg
Onions, small, minced	2	2
Celery, minced	8	8
Red bell peppers, minced	2	2
Yellow bell peppers, minced	2	2
Eggs	4	4
Cornmeal	6 oz	170 g
All-purpose flour	1 lb 4 oz	567 g
Baking powder	3 Tbsp	45 mL
Milk	8 oz	240 mL
Hot sauce	2 Tbsp	30 mL
Chopped parsley	1 cup	240 mL
Ground celery seed	2 tsp	10 mL
Dry mustard	1 Tbsp	15 mL
Garlic powder	1 Tbsp	15 mL
Paprika	2 Tbsp	30 mL
Kosher salt	as needed	as needed
Ground black pepper	as needed	as needed
Lime, sliced in wedges	as needed	as needed
Frying oil	as needed	as needed

METHOD

1. Combine all of the ingredients together and mix well.
2. Form the mixture into balls weighing 2 oz/57 g each.
3. Heat the oil to 350°F/177°C. Deep fry the fritters until fully cooked and golden brown, 2 to 3 minutes.
4. Serve with fresh lime wedges.

CRAB CAKES WITH CREOLE HONEY-MUSTARD SAUCE

YIELD: ABOUT TWELVE 4 oz/113 g CRAB CAKES

Ingredient	US	Metric
Lump crabmeat, picked clean	2 lb	907 g
Bacon, cooked crisp and crumbled	4 strips	4 strips
Fresh white bread crumbs	6 oz	170 g
Celery, small dice	6 oz	170 g
Green onions, minced	4	4
Garlic cloves, minced	4	4
Dijon mustard	1 Tbsp	15 mL
Kosher salt	as needed	as needed
Lemon juice	3/4 cup	180 mL
Cayenne	as needed	as needed
Mayonnaise	as needed	as needed
Asian-style bread crumbs (panko), for breading	10 oz	283 g
Vegetable oil	1 pt	480 mL
Creole Honey-Mustard Sauce (page 281)	1 pt	480 mL

METHOD

1. Combine the crabmeat, bacon, white bread crumbs, celery, green onions, garlic, mustard, lemon juice, salt, cayenne, and mayonnaise in a large mixing bowl and mix until homogenous; use just enough mayonnaise to hold the mixture together. Adjust the seasoning with salt and pepper.
2. Portion the crab cakes into balls weighing 4 oz/113 g each, flatten slightly, and roll in the Asia-style bread crumbs to bread them. At this point, the cakes may be refrigerated or frozen for later use.
3. Heat the oil to 350°F/177°C in a large skillet. Pan fry the crab cakes in oil until golden brown and cooked through, about 3 minutes per side. Drain briefly on paper towels. Serve immediately with the honey-mustard sauce.

Continued on next page...

FIGURE **A.2a** Pan frying crab cakes.

FIGURE **A.2b** Crab Cakes with Creole Honey Mustard Sauce.

CREOLE HONEY-MUSTARD SAUCE

YIELD: 1 qt/960 mL

Vegetable oil	1 Tbsp	15 mL
Minced shallots	1 oz	28 g
Green peppercorns, crushed	3/4 oz	21 g
Dry white wine	3/4 cup	180 mL
Ground black pepper	1 Tbsp	15 mL
Dijon mustard	2 oz	57 g
Creole mustard	6 oz	170 g
Mayonnaise	1 cup	240 mL
Sour cream	8 1/2 oz	241 g
Honey	3 Tbsp	45 mL
Kosher salt	as needed	as needed

METHOD
1. Heat the oil in a sauté pan, and sweat the shallots and peppercorns over medium heat until the shallots are translucent, 3 to 5 minutes. Do not brown.
2. Add the white wine and reduce until the liquid is almost completely evaporated, 3 to 5 minutes.
3. Transfer the mixture to a mixing bowl and let cool to room temperature.
4. Add the remaining ingredients, mix well and adjust the seasoning with salt and pepper. Refrigerate immediately and reserve until needed. Use within 3 days.

GARLIC SHRIMP

YIELD: 10 APPETIZER PORTIONS

Shrimp (21/25 count)	2 lb	907 g
Extra virgin olive oil	1 cup	240 mL
Garlic cloves, sliced	6	6
Kosher salt	4 tsp	20 mL
Red pepper flakes (optional)	as needed	as needed
Lemon juice	2 Tbsp	30 mL
Finely chopped flat-leaf parsley	2 Tbsp	30 mL

METHOD

1. Peel and devein the shrimp; reserve in the refrigerator until needed.
2. Heat the oil in a small casserole dish. Add the garlic and cook on low heat until soft but not brown, about 5 minutes. Add the shrimp, salt, and red pepper flakes, if using. Cook until the shrimp are cooked, 2 to 3 minutes.
3. Serve the shrimp sizzling, garnished with lemon juice and chopped parsley.

NOTE: Garlic shrimp should be served in small tapas portions or family style—traditionally they are served in the casserole in which they were cooked.

CLAM SAUCE

YIELD: 10 PORTIONS

Extra virgin olive oil	1 cup	240 mL
Garlic cloves, mashed	6	6
Finely chopped shallots	4 oz	113 g
Red pepper flakes	2 tsp	10 mL
Cockles or Manila clams, cleaned	80	80
Dry white wine	10 Tbsp	150 mL
Minced flat-leaf parsley	1/2 cup	120 mL
Bay leaf	2	2
Clam juice	3–5 cups	720 mL–1.2 L
Grated Parmesan cheese	1 cup	240 mL
Kosher salt	as needed	as needed
Ground black pepper	as needed	as needed

METHOD

1. Heat the oil in a pot over medium-low heat and slowly sweat the garlic, shallots, and crushed pepper until soft, about 5 minutes. Increase the heat to medium, add the clams, and briefly cook.
2. Add the wine and cook until mostly evaporated. Reduce the heat, and add the parsley and bay leaf. Cook, stirring continuously, until the clams are all opened, then transfer the clams to a warm platter as they open. Add clam juice as needed, depending on the amount of liquid left after the clams have opened. The pan should not go dry.
3. Return all of the clams to the pan, and add the cheese. Adjust the seasoning with salt and pepper.

NOTE: Toss the sauce with linguini or serve as an appetizer.

EGG PASTA

YIELD: 1 1/2 lb/680 g

Durum flour	1 lb	454 g
Kosher salt	pinch	pinch
Eggs	4	4
Oil (optional)	2 Tbsp	30 mL

METHOD

1. Combine the flour and salt in a food processor. Add the eggs and oil, if using. Process the mixture until it resembles coarse meal. When pressed, the dough will form a cohesive mass.
2. Turn the dough out onto a work surface and knead until the dough is very firm, yet pliable, adding a few drops of water if needed. Cover and let the dough relax at room temperature for at least 30 minutes.
3. Using a pasta machine or a rolling pin, roll the pasta dough into thin sheets; cut into desired shape. The pasta is ready to cook now, or it may be held under refrigeration for up to 2 days.
4. To cook, bring 1-1/2 gal/5.76 L of salted water to a rolling boil. Add the pasta and stir to separate the strands or shapes. Cook the pasta until al dente, 3 to 5 minutes, depending on the shape and size. Drain in a colander.

SHRIMP TEMPURA

YIELD: 2 lb/907 g

Ingredient	US	Metric
Pastry or cake flour	14 oz	397 g
Baking powder	1/2 tsp	2.5 mL
Egg yolks	2	2
Water, ice cold	1 3/4 cup	420 mL
Shrimp (21/25 count), peeled and butterflied	2 lb	907 g
Light sesame or vegetable oil for frying	as needed	as needed
Tempura Dipping Sauce (page 286)	1 cup	240 mL

METHOD

1. Sift together the flour and baking powder.
2. Mix the egg yolks and water into the flour, mixing just to combine; some lumps are acceptable.
3. Refrigerate the batter for 30 minutes.
4. Heat the oil to 350°F/177°C. Dip the shrimp in the batter and add to the oil. Fry until cooked through, about 1 minute. Remove the shrimp from the oil and drain briefly on paper towels before serving.

NOTES: The colder the ingredients, the crispier the batter will be.

In Japan, the most senior chef presides over the tempura station. Although at first glance it appears to be simple, the technique involves frying a variety of seafood, vegetables, and meats into light colored sesame oil. The crispy texture is achieved by using batter of the correct consistency and temperature. The item should be dropped into the oil away from you. Be sure to leave a tail of dough, which can be mounded back on the food as it cooks. Small-textured bits of dough should erupt from all sides, adding to its crisp texture. It is not unusual to have the chef deliver the food personally. Plate covers should be avoided at all costs.

TEMPURA DIPPING SAUCE

YIELD: 1 cup/240 mL

Dashi (page 294)	1/2 cup	120 mL
Mirin	1/4 cup	60 mL
Light soy sauce	1/4 cup	60 mL

METHOD

Combine all of the ingredients together and serve with tempura.

NOTE: This recipe uses a ratio of 2 parts dashi to 1 part mirin to 1 part light soy sauce.

SUSHI RICE

YIELD: 6 TO 8 PORTIONS AS PART OF A MULTI-COURSE MEAL

Short-grain rice, washed gently until water runs semi-clear	2 cups	480 mL
Water	2 1/4 cups	540 mL
Kombu (dried kelp)	3 in square	7.62 cm square
Dressing:		
Sugar	5 Tbsp	75 mL
Sea salt	4 tsp	20 mL
Unseasoned Japanese rice vinegar	5 Tbsp	75 mL

METHOD

1. Soak the rice in cold water for 30 minutes. Drain and rinse until the water runs clear and the excess starch has been removed.
2. Put the rice in a heavy-bottomed, medium-size pot or rice cooker, and add the measured water.
3. Wipe the kombu clean with a damp cloth and score it to release the flavors. Place it on top of the rice in the water.
4. Cover and place over medium heat, or turn on the rice cooker. Just before the water boils, remove the kelp and discard. Cover tightly and bring to boil over high heat for 2 minutes. Reduce the heat to medium and cook for 5 minutes. Reduce heat to very low and cook for 15 minutes, or until all of the water has been absorbed. Remove from the heat and let stand, with the pot lid wrapped in a kitchen towel, 10 to 15 minutes.
5. While the rice is cooking, prepare the vinegar dressing. In a small saucepan over low heat, dissolve the sugar and salt in the vinegar. Cool to room temperature.
6. Using a flat wooden spoon or proper rice paddle, spread the hot rice in a thin layer in a wide and shallow wooden or metal bowl, a convenient substitute for a hangiri tub. To keep the grains separate, toss the rice with horizontal, cutting strokes. This lateral motion will also keep the grains from being bruised or mashed. While tossing, sprinkle the vinegar dressing generously over rice. You may not have to use all the dressing. Be careful not to add so much liquid that the rice becomes mushy.
7. While tossing rice, cool it quickly and thoroughly with a fan.
8. To keep vinegar rice from drying out when it has cooled to room temperature, cover with a damp cloth.

NOTE: The flavor of sushi rice varies somewhat with the seasons. In summer a little more vinegar is used. Adjust the flavor of the rice as you like.

JAPANESE HAND VINEGAR

YIELD: 1 cup/240 mL

Water, cold	1 cup	240 mL
Unseasoned Japanese rice vinegar or 1 sliced lemon	1 Tbsp	15 mL

METHOD
Combine the ingredients and reserve in an accessible place.

NOTE: This mixture is used to prevent the rice from sticking to your hands. The hands are dipped in the water, then slapped together—an effective and theatrical way to remove the excess water.

WASABI

YIELD: 1/4 cup/60 mL

Wasabi powder	3 oz	85 g
Water, warm (110°F/43°C)	as needed	as needed

METHOD
Place the wasabi powder in a small bowl. Add enough warm water to achieve a smooth paste. Spread the mixture across the bottom of a bowl; wrap tightly with plastic wrap.

NOTE: Most chefs use the dry powdered form, which is not true wasabi. True wasabi is difficult to grow and expensive but can be purchased from specialty vendors.

NIGIRI SUSHI

YIELD: 30 TO 35 PIECES

Hand Vinegar (page 288)	1 cup	240 mL
Sushi Rice, cooked (page 287)	6 cups	1.44 L
Wasabi (page 288)	1 cup	240 mL

Suitable toppings for Nigiri sushi. The fish should be cut into 1 inch x 3 inch x 1/4 inch slices. Shellfish can be used whole if they are an appropriate size:

Omelet (Tamago)
Abalone
Eel
Tuna
Clam
Salmon roe
Sea urchin roe
Swordfish
Sea bream
Flounder
Sea bass
Shrimp
Mackerel

METHOD

1. Dip fingers into "hand vinegar" and rub palms together. Pick up about 1 1/2 tbsp/22.5 mL sushi rice and shape into a roughly rectangular form (or "finger") about 3 by 1 inch/7.5 x 2.5 cm.
2. Place the rice across the first joint of the fingers of one hand (the right hand for right-handed people) and form roughly by clenching that hand. With the index and middle fingers of the right hand, press and form the rice into a more defined and firm shape, turning the rice over so that all sides receive equal pressure. Do not squash or mash the rice. Smear a dab of grated wasabi horseradish or wasabi paste in the center of a slice of fish and press fish and rice "finger" together. The fish should cover the top of the "finger."

NOTES: To serve, put a small dish of soy sauce on the tray for dipping to offer along with the sushi. Sushi should be picked up with the fingers and the fish side dipped in soy sauce before being placed in the mouth with the fish side to the tongue.

Freshwater fish, and certain species of saltwater fish such as herring and salmon, all contain possible parasites and should not be eaten raw.

JAPANESE OMELET FOR NIGIRI SUSHI (TAMAGO)

YIELD: 6 OMELETS

Eggs	6	6
Dashi stock	1/4 cup	60 mL
Mirin	1 Tbsp	15 mL
Light soy sauce	1 Tbsp	15 mL
Kosher salt	pinch	pinch
Vegetable oil	as needed	as needed

METHOD

1. Whisk the eggs, dashi, mirin, soy sauce, and salt together until smooth.
2. Heat a square pan over medium heat, and then coat lightly with oil.
3. Ladle 2 oz/57 g of egg mixture into the pan, swirling it rapidly to coat evenly. Cook until the egg sets, 2 to 3 minutes. Fold the omelet away from front toward the handle in thirds.
4. Slide folded omelet away from handle to front of pan and remove from the pan.
5. Repeat this procedure until all of the egg mixture is used up.
6. Transfer omelet onto a bamboo sushi mat lined with plastic wrap. Fold the mat to form an even rectangle shape.
7. Use as a topping for nigiri sushi or in the middle of maki sushi.

NORI-ROLL SUSHI (MAKI SUSHI)

YIELD: 12 ROLLS

Nori seaweed sheets	12	12
Hand Vinegar (page 288)	3/4 cup	180 mL
Sushi Rice (page 287)	1 recipe	1 recipe
Wasabi (page 288)	3/4 cup	180 mL

Suitable Vegetable Fillings for Maki

Cucumber

Avocado

Green onion

Yam

Asparagus

Radish

Suitable Protein Fillings for Maki

Tuna

Surimi

Roe

Eel

Squid

Clams

Abalone

Octopus

Shrimp

Continued on next page...

METHOD

1. Just before rolling the sushi, toast the nori seaweed. Using whole sheets or half sheets will determine the size of the final roll. Always toast the nori sheets by passing them over a flame very briefly. Place the sheet of nori, shiny-side down, on plastic-wrapped bamboo mat.
2. Moisten your hands with the "hand vinegar" and spread about 1 to 1 1/2 cups/240 to 360 mL of sushi rice over the sheet of nori, leaving about 1/2 in/1 cm of nori showing on top and on bottom. Rice thickness should be between 1/8 and 1/4 in/3 mm and 6 mm, but do not pack it down because the rice will be compressed when it is rolled.
3. With your index finger, smear a thin line of wasabi across rice. Lay the fillings along the wasabi, making sure that the distribution is even so that roll will not be lumpy.
4. Roll ingredients in the bamboo mat as illustrated by chef (page 247–248).
5. Cut the roll in half and align the cut halves parallel to each other. Cut those halves in half—now yielding four pieces. Do not rearrange yet; cut those 2 groups of rolls in two—there will be eight pieces. Dividing into eighths creates the most common size.

NOTE: A wide range of vegetables, fish, and shellfish can be rolled into maki sushi. Flavor combinations that contrast and support each other can be used to create simple but unique parings.

MISO SOUP

YIELD: 1 gal/3.84 L

Dashi stock (page 294)	1 gal	3.84 L
Miso	1 cup	240 mL
Firm tofu, small dice	1 lb	454 g
Wakame seaweed, soaked in warm water until soft, chopped	1/2 oz	14 g
Green onions, sliced thinly on bias	1/2 bunch	1/2 bunch

METHOD

1. Place the dashi in a stockpot, and add the miso. Bring the mixture to a simmer but do not boil.
2. Add the tofu and simmer for 1 minute. Add the wakame seaweed and green onion.

NOTE: Japanese food is very seasonal, so red miso is typically used in the warmer weather whereas white miso is used in the cold weather. Boiling the dashi will greatly diminish the delicate flavors of this soup.

DASHI

YIELD: 1 gal/3.84 L

Water	5 qt	4.8 L
Kombu 2 by 3 by 6 inches	1	1
Dried bonito flakes	10 Tbsp	150 mL

METHOD
1. Fill a stockpot with the water. Crack the kombu into the water and bring to just under a simmer (185° to 195°F/85° to 90°C). Remove the kombu.
2. Keep the water at 195° to 205°F/90° to 95°C and add bonito flakes. Allow the flakes to be slowly absorbed by the water, and after about 5 minutes, carefully ladle the stock into a coffee-filter-lined strainer. The finished stock should be slightly smoky, clean tasting, and very clear.

SHRIMP BISQUE

YIELD: 1 gal/3.84 L

Shrimp shells	2 lb	907 g
Clarified butter	3/4 cup	180 mL
Onions, minced	8 oz	227 g
Garlic cloves, minced to a paste	2	2
Tomato paste	1 1/2 oz	43 g
Paprika	1 Tbsp	15 mL
Brandy	1/4 cup	60 mL
All-purpose flour	6 oz	170 g
Shellfish stock or fish stock, hot	3 qt	2.88 L
Cooked shrimp, diced	1 1/2 lb	680 g
Heavy cream	3 cups	720 mL
Kosher salt	as needed	as needed
Ground black pepper	as needed	as needed
Worcestershire sauce (optional)	1 tsp	5 mL
Cayenne pepper or Tabasco	as needed	as needed
Dry sherry	3 fl oz	90 mL

METHOD

1. Rinse the shrimp shells thoroughly in cold water and drain them.
2. Heat the clarified butter in a soup pot over medium-high heat. Add the shells and cook until they are bright red, about 10 minutes. Add the onions and cook over medium heat, stirring occasionally, until they are light brown, 5 to 7 minutes. Add the garlic and continue to cook until soft but not brown, about 3 minutes.
3. Add the tomato paste and paprika and cook over medium heat, stirring occasionally, until the tomato paste turns a rust color, 3 to 4 minutes. Add the brandy and stir well to deglaze the pan. Continue to cook until the brandy is almost cooked away, about 3 minutes.
4. Add the flour to make a blond roux and continue to cook over medium heat, stirring frequently, for 6 to 8 minutes.
5. Gradually add the stock, whisking constantly to work out any lumps. Bring to a boil, and then reduce the heat to a simmer. Simmer for 45 minutes, skimming the surface occasionally to remove any residual fat.
6. Purée the entire soup until smooth with a blender or immersion blender (shells and all). Strain the puréed soup through a fine wire-mesh sieve.
7. Bring the soup to a simmer over medium-high heat, and add the cooked shrimp and cream. Season with salt, pepper, Worcestershire sauce, if using, and cayenne. Finish with sherry and serve.

NEW ENGLAND CLAM CHOWDER

YIELD: 1 gal/3.84 L

Ingredient	US	Metric
Chowder clams	30	30
Fish stock, or as needed	2 qt	1.92 L
Salt pork, minced to a paste	4 oz	113 g
Onions, minced	4 oz	113 g
Celery, brunoise	4 oz	113 g
All-purpose flour	4 oz	113 g
Potatoes, small-dice	12 oz	340 g
Heavy cream, scalded	1 qt	960 mL
Kosher salt	as needed	as needed
Ground white pepper	as needed	as needed
Tabasco sauce	1/4 tsp	1.25 mL
Worcestershire sauce	1/4 tsp	1.25 mL

METHOD

1. Steam the clams in the stock in a covered pot over medium heat until they open, 5 to 7 minutes.
2. Strain the broth through a coffee filter or cheesecloth, and reserve. Pick the meat from the clam shells; chop and reserve the meat.
3. Render the salt pork. Add the onions and celery, and sweat until they are translucent, 6 to 7 minutes.
4. Add the flour and cook 5 to 6 minutes to make a blond roux.
5. Combine the reserved clam broth and enough additional stock to make 2 quarts/1.92 L. Gradually add to the roux and whisk to incorporate completely, working out any lumps.
6. Simmer for 30 minutes, skimming the surface as necessary.
7. Add the potatoes and simmer until tender, about 15 minutes. The soup is ready to finish now, or it may be rapidly cooled and stored for later service.
8. Return the soup to a simmer over low heat. Add the reserved clams and cream. Adjust seasoning with salt, pepper, Tabasco, and Worcestershire.

MUSSELS MARINIÈRE

YIELD: 10 PORTIONS

Unsalted butter	8 oz	227 g
Minced shallots	1 1/2 cups	360 mL
Mussels, de-bearded	8 qt	7.68 L
White wine	3 1/2 cups	840 mL
Fish stock	1/2 cup	120 mL
Lemon juice	1/2 cup	120 mL
Kosher salt	as needed	as needed
Ground black pepper	as needed	as needed
Chopped parsley	1/2 cup	120 mL
Chopped thyme	1/4 cup	60 mL
French bread	as needed	as needed

METHOD

1. In a large pot, melt 4 oz/113 g of the butter over medium heat and sweat the shallots until tender but not brown, about 5 minutes.
2. Add the mussels, white wine, and fish stock. Cover and cook until the mussels are open, 5 to 7 minutes.
3. Remove the mussels from the pot, discarding any mussels that did not open. Add the lemon juice to the liquid in the pot. Reduce the liquid by about one-third, about 8 minutes, and then reduce the heat to low.
4. Season the liquid with salt and pepper and whisk in the remaining butter.
5. Finish the sauce with parsley and thyme.
6. Divide the mussels among 10 bowls and pour the sauce over them.
7. Serve with crusty French bread.

BOUILLABAISSE

YIELD: 20 PORTIONS

Olive oil	1/2 cup	120 mL
Onions, sliced	1 lb	454 g
Leeks, sliced	1 lb	454 g
Fennel, sliced	8 oz	227 g
Garlic cloves, chopped	8	8
Tomatoes, peeled and seeded or whole canned, chopped	2 lb	907 g
Tomato paste	2 Tbsp	30 mL
Crumbled saffron threads	2 tsp	10 mL
Bouquet garni	1	1
Potatoes, peeled, cut into 1/2 in dice	1 lb	454 g
Fish stock	3 qt	2.88 L
Kosher salt	as needed	as needed
Ground black pepper	as needed	as needed
Cayenne pepper	as needed	as needed
Assorted seafood (John Dory, Red mullet, Red snapper, Porgy, Monkfish, Lobster), cut into 3 in/7.5 cm chunks	3 lb	1.36 kg
Chopped parsley	1/2 cup	120 mL
Chopped chervil	1/2 cup	120 mL
Baguette or Country bread loaf, cut into slices, dried in oven	1	1
Rouille (page 301)	as needed	as needed

METHOD

1. In large, heavy-bottomed stockpot, heat the olive oil over medium high heat and sweat the onions, leeks, fennel, and garlic until translucent, about 5 minutes.
2. Add the tomatoes, tomato paste, saffron, bouquet garni, potatoes, and fish stock and simmer until the potatoes are cooked, 10 to 15 minutes. Season with salt, black pepper, and cayenne pepper.

Continued on next page...

3. Add the seafood in order of firmness and cooking time, with the denser, longer cooking items placed on the bottom, and the delicate fillets on top. Cook until the seafood is done. Once the seafood is cooked, add the parsley and chervil.
4. To serve, carefully remove fish and shellfish, arranging it on a large serving platter or in a warm bowl. Ladle the broth and potatoes over fish, and serve with the bread and Rouille.

NOTE: Bouillabaisse is traditionally served with toasted French bread and Rouille, a garlic and saffron mayonnaise.

FIGURE A.3 Bouillabaisse

ROUILLE (GARLIC AND SAFFRON MAYONNAISE)

YIELD: 8 oz/227 g

Garlic cloves	3	3
Kosher salt	1/4 tsp	1.25 mL
Cayenne pepper	1/4 tsp	1.25 mL
Saffron, powdered, dissolved in 1 Tbsp/15 mL boiling water	1/2 tsp	2.5 mL
Fresh bread crumbs	1/4 tsp	1.25 mL
Egg yolk, room temperature	1	1
or pasteurized egg yolk	1 Tbsp	15 mL
Lemon juice	1 tsp	5 mL
Olive oil, room temperature	1 cup	240 mL

METHOD

1. In a food processor or mortar and pestle, process the garlic, salt, and cayenne pepper to form a paste. Mix in the dissolved saffron and bread crumbs.
2. Add the egg yolk and lemon juice and combine thoroughly.
3. Add the oil in a slow steady stream; process until the sauce emulsifies and thickens. Reserve in the refrigerator until needed.

SALADE NIÇOISE

YIELD: 10 PORTIONS

Roasted New Potatoes		
New potatoes, washed, skin on	3 lb	1.36 kg
Pure olive oil	6 Tbsp	90 mL
Garlic cloves	8	8
Thyme sprigs	1/4 bunch	1/4 bunch
Bay leaves	2	2
Kosher salt	1 1/2 tsp	7.5 mL
Ground black pepper	1 tsp	5 mL
Green beans	1 1/2 lb	680 g
Tuna fillets (3 oz/85 g each)	10	10
Olive oil	as needed	as needed
Kosher salt	as needed	as needed
Ground black pepper	as needed	as needed
Red onion, julienne	3 oz	85 g
Chopped flat-leaf parsley	1/4 cup	60 mL
Capers, rinsed and chopped	1/4 cup	60 mL
Red Wine Vinaigrette	1 cup	240 mL
Mixed greens	10 oz	284 g
Roma tomatoes, peeled, quartered	5	5
Eggs, hard-boiled, peeled, quartered	5	5
Anchovy fillets	10	10
Niçoise or kalamata olives	40	40

METHOD

1. For the roasted potatoes: Toss the potatoes with the olive oil, garlic, thyme, bay leaves, salt, and pepper in a large bowl. Roast in a 355°F/180°C oven until fork-tender. Remove from the oven and reserve until needed.
2. Blanch the green beans in salted boiling water until tender, about 5 minutes, and then shock in ice water until they stop cooking. Reserve until needed.
3. Lightly brush the tuna fillets with olive oil and season with salt and pepper. Grill over medium-high heat until the tuna is medium-rare, 2 to 3 minutes per side. The fish will still appear translucent in the center. Remove and keep warm.

Continued on next page...

4. Toss the roasted potatoes, green beans, onion, parsley, and capers together. Add the vinaigrette and toss to coat.
5. Arrange the greens on a platter. Place the cooked tuna in the center. Arrange the beans and potato mixture around the tuna. Garnish the plate with the tomatoes, eggs, anchovies, and olives.

FIGURE A.4 Salade Niçoise

TROUT WITH SAUTÈED MUSHROOMS

YIELD: 10 PORTIONS

Milk	1/2 cup	120 mL
Rainbow or brook trout, cleaned (about 8 oz/227 g each)	10	10
Kosher salt	as needed	as needed
Ground black pepper	as needed	as needed
All-purpose flour	8 oz	227 g
Clarified butter	1/2 cup	120 mL
Butter, whole	8 oz	227 g
Assorted mushrooms, cleaned, cut into bite-sized pieces	2 lb	907 g
Minced shallots	4 oz	113 g
Capers	6 oz	170 g
Lemons, juice only	2	2
Chopped thyme	1 Tbsp	15 mL
Minced chives	1 Tbsp	15 mL

METHOD

1. Pour the milk into a flat, shallow half hotel pan. Dip the fish in the milk, turning several times to coat well. Season the trout with salt and pepper.
2. Pour the flour into a second flat, shallow hotel pan and coat the fish with the flour, shaking to remove any excess.
3. Heat the clarified butter over medium heat in a sauté pan large enough to hold the trout in a single layer. This may need to be done in batches. Add the trout and cook on the first side for about 4 minutes. Turn and cook on the second side for about 4 minutes, basting frequently with the clarified butter so that they brown evenly. Transfer the fish to a heated serving platter and keep warm. Discard any oil remaining in the skillet.
4. In the same skillet, heat 4 oz/113 g of the butter over high heat. Add the mushrooms, and season with salt and pepper. Cook, tossing, until lightly browned, 5 to 10 minutes. Add the shallots and capers, and cook briefly, 2 to 3 minutes. Add the lemon juice and stir in the remaining butter. Adjust seasoning as needed, and add the thyme and chives.
5. Spoon the mushroom mixture over the trout and serve.

DOVER SOLE MEUNIÈRE

YIELD: 1 PORTION

Dover sole, skinned and trimmed	1	1
Lemon juice	as needed	as needed
Kosher salt	1/4 tsp	1.25 mL
Ground black pepper	pinch	pinch
All-purpose flour, for dredging	as needed	
Clarified butter	1 oz	28 g
Lemon juice	1 Tbsp	15 mL
Chopped parsley	1 Tbsp	15 mL
Butter, whole	1/2 oz	14 g

METHOD

1. Remove the skin from the Dover sole by pulling it off from the tail end (page 170). Cut off the fins, and if desired, the head. Season with lemon juice, salt, and pepper; dredge in flour.
2. Sauté the fish in a large sauté pan in the clarified butter over medium heat until lightly browned and cooked through, 4 to 5 minutes per side.
3. Transfer to a serving platter and season with the lemon juice and parsley.
4. Wipe out the pan and add the butter. Heat the butter until lightly browned, 2 to 3 minutes, and pour over the fish.

NOTE: Dover Sole is classically served whole and filleted tableside. Steamed potatoes are an excellent accompaniment.

Continued on next page...

FIGURE **A.5a** Sautéing Dover Sole.

FIGURE **A.5b** Dover Sole Meunière.

SHRIMP WITH TOMATOES, FETA, AND OREGANO

YIELD: 10 PORTIONS

Ingredient	US	Metric
Shrimp (16/20 count), shelled and deveined	3 lb	1.36 kg
Kosher salt	as needed	as needed
Ground black pepper	as needed	as needed
Olive oil	1/2 cup	120 mL
Shallots, sliced	4 oz	113 g
Garlic cloves, minced	8	8
Oregano	1/4 cup	60 mL
Cayenne pepper	1/2 tsp	2.5 mL
Tomatoes, peeled and seeded or canned tomatoes, chopped	1 lb	454 g
Feta cheese, crumbled	8 oz	227 g
Chopped flat-leaf parsley	1/4 cup	60 mL
Toasted French bread	as needed	as needed

METHOD

1. Season the shrimp with salt and pepper. Briefly sauté the shrimp in 2 tablespoons of the oil over high until light brown but still raw in the middle, 1 minute.
2. Transfer the shrimp to a baking dish.
3. Heat the remaining oil in the sauté pan and sweat the shallots until translucent, about 2 minutes. Add the garlic, oregano, and cayenne; sauté for 2 minutes or until the garlic softens and browns. Add the tomatoes and simmer for a few minutes; season with pepper. Pour the sauce over shrimp.
4. Sprinkle the crumbled cheese over sauce. Broil or bake until cheese just begins to brown, about 3 minutes.
5. Sprinkle with chopped parsley and serve with crusty bread.

PAELLA

YIELD: 6 PORTIONS

Ingredient	US	Metric
Chicken or fish stock, well seasoned	5 cups	1.2 L
Saffron	1 Tbsp	15 mL
Kosher salt	as needed	as needed
Olive oil	6 Tbsp	90 mL
Chicken thighs, bone in	1 lb	454 g
Spanish chorizo, cut in half-moons	8 oz	227 g
Onion, medium dice	2 oz	57 g
Spanish paprika	1 Tbsp	15 mL
Minced garlic	1 Tbsp	15 mL
Red bell pepper, medium dice	6 oz	170 g
Plum tomatoes, canned	3 oz	85 g
Short grain Spanish rice	2 1/2 cups	600 mL
Mussels	12	12
Clams	6	6
Shrimp (21/25 count)	12	12
Green peas, blanched	3 oz	85 g
Lemons, halved	2	2

METHOD

1. Warm the saffron in the stock for about 5 minutes to infuse it. Bring the stock to a boil and season with salt. The stock should taste slightly overseasoned to avoid stirring in salt once the paella is ready to serve.
2. Heat the olive oil and sauté the chicken and chorizo over medium high heat in a paella pan, about 8 minutes. When the chicken and chorizo are fully cooked, remove them from the pan and keep warm.
3. Add the onions, paprika, garlic, peppers, and tomatoes, and cook until very soft about 5 minutes.
4. Increase the heat to high and add the rice. Cook the rice, stirring frequently, for 3 to 4 minutes to break down the outer starch layer. Return the chicken and chorizo to the pan and pour in the hot stock, bringing it to a boil.
5. Reduce the heat, cover the pan and simmer for about 10 minutes. Remove the lid and arrange the mussels, clams, and shrimp over the rice and continue to cook for 5 to 8 additional minutes or until the clams and mussels open and the shrimp is fully cooked.

Continued on next page...

6. Remove from the heat and let rest for 5 minutes.
7. Garnish with peas and squeeze the lemon over the rice and seafood. Serve the paella in the paella pan.

NOTE: The liquid ratio may vary depending on the type of rice used, so read the package instructions. Paella is typically made with a short-grain Spanish rice and served directly in the pan.

FIGURE **A.6a** Adding the rice to the cooked onions, paprika, garlic, peppers, and tomatoes.

FIGURE **A.6b** Adding the hot saffron-infused broth.

FIGURE **A.6c** Arranging the shrimp over the rice.

FIGURE **A.6d** Paella

BASE RECIPE FOR SHALLOW POACHED FISH

YIELD: 10 PORTIONS

Fish fillets, portioned and seasoned	3-3/4 lb	1.7 kg
Butter, whole	3 oz	85 g
Minced shallots	4 oz	113 g
White wine, dry	4 oz	113 g
Fish stock	8 oz	227 g

METHOD
1. Butter the bottom of a high-sided sauté pan and sprinkle the bottom with the shallots.
2. Place the portioned fish on the shallots.
3. Add the wine and fish stock to the pan.
4. Cover with a buttered parchment paper round the size of the sauté pan, buttered side down.
5. On the stove top, bring the mixture to a simmer over medium heat, and then place into a preheated 325°F/163°C oven.
6. Cooking times will vary based on thickness of the portion, but range between 5 and 15 minutes.
7. Carefully remove the fish and place on a warm plate with paper toweling to absorb excess liquid; cover with the buttered paper from the poaching process or plastic wrap.
8. Reduce the cooking liquid to a few ounces over high heat.
9. Prepare your sauce from the variations below.

Sauce variations:

Beurre Blanc:
1. Reduce the cooking liquid (cuisson), and whip in enough whole butter to establish a sauce consistency.
2. Add heavy cream, reduce until thickened and finish with whole butter.

Vin Blanc:
1. Add pre-made velouté to the reduction of the cuisson, and finish with cream or monter au beurre.

Continued on next page...

Classical Garnishes for Poached Fish
Bercy: parsley, butter
Dugleré: tomato concassé, parsley, velouté, butter
Dieppoise: mussels, shrimp, velouté, liaison

FIGURE A.7a
Shallow poaching fish fillets.

FIGURE A.7b
Shallow Poached Fish with Sauce Vin Blanc

FILLET OF FLOUNDER WITH WHITE WINE SAUCE

YIELD: 10 PORTIONS

Flounder fillet, trimmed (5 oz/142 g)	10	10
Kosher salt	2 tsp	10 mL
Ground black pepper	1/2 tsp	2.5 mL
Butter	1/4 cup	60 mL
Minced shallots	5 Tbsp	75 mL
Dry white wine	1 pt	480 mL
Fish stock	3 cups	720 mL
Heavy cream	1 pt	480 mL
Chopped parsley	1/4 cup	60 mL
Chopped chervil	1/4 cup	60 mL

METHOD

1. Season the flounder with salt and pepper, and roll it tail to head with the skin side in. Reserve until needed.
2. Butter an ovenproof pan and sprinkle with the shallots. Place the flounder in the pan, and add the wine and stock.
3. Bring the liquid in the pan to a simmer on the stovetop over medium heat.
4. Cover the fish with buttered parchment, or waxed paper (a cartouche), and transfer to a 350°F/177°C oven. Continue cooking until the flesh turns opaque, 5 to 7 minutes.
5. Remove the flounder from the pan and keep it warm until ready to serve.
6. Reduce the poaching liquid by two thirds over high heat. Add the heavy cream and reduce the sauce until it coats the back of a spoon, 5 to 7 minutes. The sauce can be strained at this point, if desired.
7. Add the chopped parsley and chervil and adjust seasoning with salt and pepper.
8. Serve the flounder accompanied with the sauce.

GRILLED SALMON WITH GINGER GLAZE

YIELD: 10 PORTIONS

Salmon fillets or steaks (6 oz/170 g each)	10	10
Kosher salt	2 tsp	10 mL
Marinade		
Soy sauce	3/4 cup	180 mL
Sake or dry white wine	3/4 cup	180 mL
Mirin wine	3/4 cup	180 mL
Honey	3 Tbsp	45 mL
Chopped green onion	1/2 cup	120 mL
Chopped ginger	1/2 cup	120 mL

METHOD

1. Season the salmon with the salt.
2. For the marinade: Combine all of the marinade ingredients and blend together well. Pour the marinade over the salmon. Refrigerate for 1 to 2 hours.
3. Remove the salmon from the marinade and grill over high heat on both sides until just done, approximately 3 to 4 minutes per side depending on the thickness of the fish. Baste the salmon with the remaining marinade as it cooks.

Continued on next page...

FIGURE **A.8a** Grilling salmon.

FIGURE **A.8b** Grilled Salmon with Ginger Glaze

PAN-FRIED COD

YIELD: 10 PORTIONS

Cod fillet, trimmed, 6 oz pieces	3 3/4 lb	1.7 kg
Kosher salt	1 tsp	5 mL
Ground black pepper	1/4 tsp	1.25 mL
Lemon juice	1/2 cup	120 mL
All-purpose flour	1 lb	454 g
Eggs, beaten	4	4
Bread crumbs	1 lb	454 g
Vegetable oil	as needed	as needed

METHOD

1. Season the cod with salt, pepper, and lemon juice.
2. Prepare a standard breading mise en place (see page 264–266). Immediately prior to pan frying, dredge the cod in flour, dip in beaten egg, and coat in bread crumbs.
3. Heat the oil in a large sauté pan over medium heat. Pan-fry the cod in hot oil until golden brown on both sides and cooked through, 3 to 4 minutes per side.
4. Remove from pan and place on a wire rack to drain excess fat.

FISH AND CHIPS

YIELD: 1 1/2 lb/680 g OF FISH

Cod	1 1/2 lb	680 g
Lemon juice	1 Tbsp	15 mL
Extra virgin olive oil	1 Tbsp	15 mL
Kosher salt	1/2 tsp	2.5 mL
Batter		
All-purpose flour	1-1/2 cups	360 mL
Baking powder	1 tsp	5 mL
Kosher salt	1/4 tsp	1.25 mL
Egg yolks	2	2
Ice water	1 cup	240 mL
Flour for dredging	as needed	as needed
Vegetable oil for frying	as needed	as needed
French Fried Potatoes (page 325)	as needed	as needed

METHOD

1. Cut the cod into portion sizes, and marinate it in lemon juice, olive oil, and salt.
2. For the batter: Mix the dry ingredients and the wet ingredients in separate bowls.
3. Pour the wet ingredients into the dry ingredients and mix until smooth. Refrigerate for 1 hour.
4. Dredge the fish in flour and then dip into the batter.
5. Heat the oil to 360°F/182°C. Deep fry the cod in the oil until golden brown and fully cooked, 2 to 4 minutes. Remove from the oil and drain briefly on paper towels to remove the excess oil. Serve with the French fried potatoes.

NOTE: To achieve a very crispy batter, it must be very cold. When cold batter is dropped into the appropriate temperature oil it will immediately begin to cook and form a crispy exterior. If the batter is warm or the oil is not the correct temperature, the product will be oily and not as crispy.

Continued on next page...

FIGURE **A.9a** Frying battered cod.

FIGURE **A.9b** Fish and Chips

FISH AND CHIPS

FRENCH FRIED POTATOES

YIELD: 10 PORTIONS

Idaho potatoes, peeled	3 1/2 lb	1.59 kg
Vegetable oil, for deep frying	as needed	as needed
Kosher salt	1 tsp	5 mL

METHOD

1. Cut the potatoes into the desired shape. Rinse in cold water, and then dry thoroughly.
2. Heat the oil to 275° to 300°F/135° to 149°C. Add the dried potatoes in batches to the oil and blanch for 2 minutes. This should be done in batches. Drain the blanched potatoes well and transfer the potatoes to pans lined with paper towels.
3. When all of the potatoes have been blanched, increase the oil temperature to 350°F/177°C.
4. Return the potatoes to the oil in batches and continue cooking until golden brown, about 3 minutes.
5. Remove from the oil and shake off excess fat. Season with salt after draining excess oil.

LOBSTER THERMIDOR

YIELD: 6 PORTIONS

Lobsters (1 1/2 lb/680 g each)	6	6
Unsalted butter	2 oz	57 g
Minced shallots	1/2 cup	120 mL
Mushrooms, small button, quartered	2-1/2 cups	600 mL
Cognac	1/2 cup	120 mL
Sherry	1/2 cup	120 mL
Heavy cream	6 cups	1.44 L
Dijon mustard	1/2 cup	120 mL
Chopped chervil	1/2 cup	120 mL
Chopped tarragon	5 Tbsp	75 mL
Grated Parmesan cheese	1 cup	240 mL
Kosher salt	as needed	as needed
Ground black pepper	as needed	as needed

METHOD

1. Boil the lobsters in a stockpot filled with water for 1 to 2 minutes. The lobsters will still be raw in the middle but easy to remove from the shell.
2. Place the lobsters on their backs and cut in half lengthwise. Remove the tail meat, cut into cubes, and reserve. Clean the bodies of viscera and save the shells for service. The roe and liver can be saved and stirred into the sauce at the end.
3. Separate the claws and knuckles from the body, crack them, remove the meat, and cut into cubes.
4. In a large sauté pan, melt the butter over low heat and sweat the shallots and mushrooms until tender but not brown, about 5 minutes.
5. Deglaze the pan with the cognac and sherry and reduce by two thirds.
6. Add the cream and reduce by at least half or until the sauce coats the back of a spoon. Stir in the mustard.
7. Add the cubed lobster meat and cook until heated through, about 3 minutes.
8. Add the chervil, tarragon, and most of the parmesan cheese. Season with salt and pepper.
9. Fill the shells with the lobster mixture and sprinkle with parmesan cheese. Broil until the cheese lightly browns, 3 to 5 minutes.

COLD-SMOKED SALMON

YIELD: 1 SIDE OF SALMON

Cure

Kosher salt	12 oz	340 g
Dark brown sugar	6 oz	170 g
Ground black pepper	2 tsp	10 mL
Ground pickling spice	2 Tbsp	30 mL
Onion powder	2 Tbsp	30 mL
Garlic powder	1 Tbsp	15 mL
Salmon fillet, scaled, skin on, boneless	1 side	1 side

METHOD

1. Line a hotel pan with plastic wrap.
2. Combine all of the cure ingredients.
3. Place 4 ounces/113 grams of the cure mixture on the bottom of the pan.
4. Place the salmon fillet, skin side down, on the cure.
5. Sprinkle 8 ounces/227 grams of cure mixture onto the top of the fish, spreading the cure slightly heavier where the fish is thicker.
6. Wrap the fillet and cure mix tightly with the plastic wrap. Use more plastic wrap as required to make a tight package.
7. Refrigerate for 20 hours, turning once halfway through the process. The fillet may be evenly weighted with a 5 lb/2.27 kg weight during this time to ensure a more uniform thickness to the product. However, this is optional.
8. After curing, rinse under cold water to remove the cure.
9. Blot dry using paper towels.
10. Air-dry to form pellicle, on a rack in the refrigerator for 1 day.
11. Place in a cold smoker and smoke for 2 to 4 hours, depending on the smoke density of the unit and the desired degree of smoke flavor.
12. Cool again for at least an hour before slicing or wrapping for storage.

GRAVLOX (DILL CURED SALMON)

YIELD: 1 SIDE OF SALMON

Gravlox Cure		
Kosher salt	4 oz	113 g
Dark brown sugar	8 oz	227 g
Ground black pepper	1 tsp	5 mL
Garlic powder	2 tsp	10 mL
Dill, chopped	1 pt	480 mL
Onion, fresh, minced (optional)	1/2 cup	120 mL
Salmon fillet, scaled, skin on, boneless	1 side	1 side
Dill, chopped	1/4 cup	60 mL

METHOD

1. Insert a perforated 2-inch hotel pan into a solid 4-inch pan and line it with cheesecloth. Combine all of the cure ingredients. Place 4 oz/113 g of the cure mixture on the bottom of the pan.
2. Place the fillet, skin side down, onto the cure.
3. Sprinkle the chopped dill over the fish and pack the remaining cure mixture on the top of the fish, spreading the cure slightly heavier where the fish is thicker.
4. Wrap tightly with the cheesecloth.
5. Refrigerate for 2 to 3 days, turning over every 12 hours. The fillet may be evenly weighted down with a 5 lb/2.27 kg weight during this time to ensure a more uniform thickness to the final product. However, this is optional.
6. After curing, gently scrape off the cure and rinse until it is completely removed. Blot dry using paper towels and sprinkle with additional chopped dill.

SALMON RILLETTE

YIELD: ABOUT 2 lb/907 g

Minced shallot	1/4 cup	60 mL
White wine	4 oz	113 g
Salmon	12 oz	340 g
Smoked salmon, small dice	12 oz	340 g
Butter, softened	4 oz	113 g
Mayonnaise	4 oz	113 g
Zest of lemon	1 tsp	5 mL
Dill, chopped	1/2 oz	14 g
Lemon juice	as needed	as needed
Kosher salt	as needed	as needed
Ground black pepper	as needed	as needed

METHOD

1. Place the shallots and white wine in a pot and bring to boil over medium heat.
2. Add the fresh salmon and smoked salmon to the pot and reduce the heat to low. Poach until just cooked and then cool. The cooked temperature of the salmon should be 145°F/63°C.
3. Save the cooking liquid and reduce until syrupy, about 5 minutes.
4. Flake the fresh salmon into small pieces.
5. Purée the smoked salmon with the butter in a food processor.
6. Transfer the puréed salmon to a mixing bowl and mix in the flaked salmon, the remaining ingredients, and some of the poaching liquid to adjust the consistency until it resembles tuna salad.
7. Adjust seasoning as needed.
8. Refrigerate for at least 4 hours before serving.

FISH STOCK

YIELD: 1 gal/3.84 L

Ingredient	US	Metric
Onions, sliced	8 oz	227 g
Celery	4 oz	113 g
Mushroom trimmings, sliced	4 oz	113 g
Parsley stems	1 oz	28 g
White-fleshed fish (bones and trimmings, no gills)	8 lb	3.63 kg
Water	5 qt	4.8 L
Dry white wine	8 fl oz	240 mL
Lemon juice	1 Tbsp	1 Tbsp
Peppercorns	10 ea	10 ea
Bay leaves	5 ea	5 ea
Thyme leaves	1 tsp	1 tsp
Kosher salt	3/4 oz	21 g

METHOD

1. Lay the onions, celery, mushrooms, and parsley stems in the bottom of a stock pot. Lay the fish bones and trim on top.
2. Cover the vegetables and fish bones with the water, wine, and lemon juice. Bring the liquid to a quick boil over medium heat and skim carefully. Reduce the heat to low and simmer for 30 minutes.
3. Add the remaining ingredients, simmer for an additional 15 minutes, and then strain through a fine mesh sieve.
4. Ice down the stock to cool quickly if not using immediately. Store, covered, in a refrigerator.

MOUSSELINE FORCEMEAT

YIELD: 1 1/2 lb/680 g

Fish or shellfish, skinless and boneless	1 lb	454 g
Kosher salt	1 tsp	5 mL
Seasoning	as required	as required
Egg whites	2	2
Heavy cream, cold	3/4 to 1 cup	180 to 240 mL
Garnishes, low moisture, soft	as required	as required

METHOD

1. Keep the fish or shellfish very cold and on ice throughout the process. Dice the fish and add salt.
2. Fine grind the fish if it has resilience, such as crustaceans, trout, or monkfish. Very soft seafood such as scallops and most fatty fish will not need to be initially ground.
3. Purée the lightly salted seafood and seasoning in a chilled food processor with the egg whites, until very smooth. Open the processor occasionally and scrape down the sides with a rubber spatula to ensure a uniformly fine result. Chill if the mixture gets warmer than 45°F/7°C.
4. Pulse in the cold cream, incorporating it fully, 10 seconds. Open the processor lid and scrape down the sides with a rubber spatula at least once during this step. Processing the cream in excess of 10 seconds may either aerate the forcemeat, causing it to rise like a soufflé, only to collapse while cooling, or churn the butterfat from the cream, causing a fat separation (breaking) during cooking. The consistency should be similar to mayonnaise.
5. Test a quenelle of forcemeat by poaching it in 180°F/82°C water for 5 minutes.
6. Adjust the seasoning as desired.
7. Fold in fresh herbs, garnishes, and/or inlays only after the base forcemeat proves successful both in texture and flavor.

GLOSSARY

Abalone—A univalve mollusk. It has a beautiful, oval shaped flat shell. Its white flesh is tough and requires tenderizing; the entirety of the flesh is edible and its shell is valued as inlay material for jewelry and furniture making.

Anchovy—Iridescent gray and blue-green colored, this fin fish averages six to nine inches in length and is prized for its preservation abilities and unique flavor. The most common species for culinary purposes is the European anchovy. Anchovies are most commonly sold in canned form or in a paste form in the U.S.; their "fishy" flavor is largely due to curing processes. Fresh anchovies, available seasonally, are rich and moist.

Algaculture—The controlled farming of seaweed.

Aquaculture—The process of gathering fish in their natural habitat and raising them within controlled environments for cultivation and harvest purposes.

Aquaponics—A fish farming practice combining aquaculture and hydroponics, where waste and water from fish serve as nutrients to plants which in turn filter water back into the tank used by the fish.

Arthropod—A very large group of invertebrates; marine crustaceans with soft bodies protected by jointed exoskeletons. Rigid shells are shed and regrown through periodic molting, and bristles on their exterior are used for sensory purposes.

Bass—A round fish with firm, smooth flesh ideal for nearly all cooking applications. Species are commonly found in salt waters of both the Atlantic and the Pacific, as well as in some fresh waters within the United States.

Bivalve—A class of mollusk with a hinged shell; the protective shell is used to either propel the invertebrates or to secure themselves from hard objects. Bivalves breathe through a sheet-like gill through which water is filtered.

Braise—A cooking method in which the main item is seared in fat, then simmered at a low temperature in a small amount of stock or another liquid (usually halfway up the item) in a covered vessel for a long time. The cooking liquid is then reduced and used as the base of a sauce.

Brine—A solution of salt, water, and seasonings, used to preserve or moisten foods.

Broil—To cook by means of a radiant heat source placed above the food.

Butterfly—A technique used on meat or seafood where an item is cut in order to spread out its edges, the symmetrical final shape of which resembles butterfly wings.

Catfish—This fast-growing fin fish is in shades of gray with black spots; it has a forked tail and characteristic whiskers. The fish may be found in freshwater lakes and farm-raised in ponds, as well as in channels running as far south from the Great Lakes as Mexico. The flesh is moist, dense, sweet, and mild.

Clam—A bivalve mollusk with over 12,000 species worldwide, it ranges in size from very tiny to over three feet in diameter. Their shells are locked together by strong adductor muscles; clams are often eaten raw as well as cooked using various preparations depending on the species.

Cephalopods—A flexible mollusk, they are known to have unusually developed vision and memory. A large bulbous mass containing the brains, mouth, and organs surrounds the rest of the animal. The invertebrate's foot is composed of numerous suction-cupped tentacles.

Cod—Found in areas where cold and warm waters converge, cod is an extremely versatile fish for cooking. It has brown spots speckling its brownish body that is lined with white stripes on either side. The majority of species live in the Northern hemisphere, the most commercially important being the Atlantic cod.

Conch—A tropical gastropod with a pinkish, spiral shell surrounding the edible muscle. Similar in flavor and texture to clams, the meat is tough and must be tenderized.

Coquilles St. Jacques—The French term for diver scallops; as a preparation, scallops served in the half shell.

Crab—Readily found around the North American continent as well as in eastern parts of Asia, it is an arthropod with eight claws and depending on the species has a hard or soft shell. While laborious to remove, crab meat is sweet and rich.

Cross-contamination—When harmful or disease causing elements are transferred from one contaminated surface to another. This can be prevented through hygienic practices such as wearing gloves.

Crustacean—A class of hard-shelled arthropods with elongated bodies, primarily aquatic, which include edible species such as lobster, crab, shrimp, and crayfish.

Cure—To preserve a food by salting, smoking, pickling, and/or drying.

Dashi—A Japanese fish stock made of water, kombu seaweed, and dried bonito flakes. It is used in a wide variety of stocks and soups in Japanese cuisine.

Deep fry—To cook food by immersion in hot fat; deep-fried foods are often coated with bread crumbs or batter before being cooked.

Deep poach—To cook food gently in enough barely simmering liquid to completely submerge the food.

Dredging—Primarily used for catching shellfish, a fishing method in which a metal basket is pulled through the seabed. The rake-like front of the cage stirs up silt, producing the life beneath it that is then scooped into the vessel.

Drift nets—Invisible to fish, a wall-like mesh net used to catch squid, tuna, salmon and other valuable species. Sometimes referred to as "ghost nets," drift nets are unattached from boats or anchors thus are easily lost and responsible for catching and damaging unintended sea life.

Dry cure—A combination of salts and spices used to preserve meats; often used before smoking to process meats and forcemeats.

Eel—Snakelike, a fish firm in texture with gray skin that turns white when cooked. They spawn in salt waters and return to fresh waters where they are caught. Eel are often sold live as they can survive out of the water for several days.

FIFO—An industry term which stands for "first in, first out," referring to the concept of using older goods first, and storing products accordingly.

Fillet—The most popular market form of fish, it is the entire side of a fish, from head to tail, removed from the back bone. Fillets may be cut into smaller portions and are generally boneless; larger cuts store and cook better than those that are smaller.

Fish traps or pots—A selective method of fishing where a trap is baited and lowered to the sea floor until the desired species swims into the vessel. Pots are buoyed and are desirable in that they deliver a catch live with little unwanted bycatch.

Flounder—A bottom-dwelling fin fish distinguishable by its compressed, boney body and both its eyes are on one side of its body. This family of fish has notably light and dark coloring on each side of its body respectively, and has the ability to change its color depending on the environment. Availability is widespread, from both U.S. coasts to Europe and Asia.

Food-borne illness—A gastrointestinal disorder caused by the consumption of food or drink that has been contaminated through improper cooking, handling, or storage.

Gastropod—A class of mollusk recognizable by its single, usually spiral shaped shell. Shells may also be flat; some have no shells at all. The foot of the shellfish is used for propulsion.

Gillnetting—A fishing method employing long walls of mesh, varying in size and invisible to fish. Once the head and gills of a fish have passed through the net, the fish becomes entangled and dies. Gill nets may be placed at various depths within the water, depending on the species and location.

Grill—A cooking technique in which foods are cooked by a radiant heat source placed below the food. Also, the piece of equipment on which grilling is done. Grills may be fueled by gas, electricity, charcoal, or wood.

Gutting—The first step of fish fabrication in which the viscera is removed from the animal. It is important to do this promptly after catching the fish as an enzyme present in the guts breaks down quickly, leading to spoilage.

Halibut—The largest of the flat fish family, it is found in deep, cold waters of the Atlantic and Pacific. Recognizable by its size as well as its contrasting white and dark sides, it has mild flavor and firm flesh; its large size makes it ideal for various specialty cuts for market sale.

Hazard Anaylsis Critical Control Point (HACCP)—A monitoring system used to track foods from the time that they are received until they are served to consumers, to ensure that the foods are free from contamination. Standards and controls are established for time and temperature, as well as for safe handling practices.

Ice—In various shapes and sizes, made from fresh or salt water, extremely effective in keeping fish fresh as well as hydrated. Often combined in slurries with water for cooling large batches of fish at once.

In the round—A market form of fish where the animal is whole and unprocessed. Head, scales and viscera remain attached.

Infection—A food-borne illness caused by the ingestion of food containing bacteria that multiplies within the body, attacking the gastrointestinal lining.

Intoxication—A food-borne illness caused by the ingestion of toxins from mold, bacteria or certain plants and animals.

Lobster—An arthropod with large claws and shells that are molted as the shellfish grows. Almost entirely edible, the red and white meat found in claws, tail, knuckles and feet is sweet and varies in firmness depending on the area from which it is taken. Lobster should be kept alive until just before preparation; the most common species is the Maine or American lobster.

Longlining—A highly productive method of catching fish in which a baited line is dropped into the water with hooks at various levels of depth. Lines are marked with buoys at the surface. Lines may be vertical or horizontal, balanced and suspended by bottom anchors and surface buoys.

Mackerel—A round fish with a soft texture and flesh that flakes when cooked; the flavor depends on spawning and feeding cycles. All varieties of mackerel should be iced quickly after capture to prevent the development of scrombroid poisoning.

Mahi-Mahi—Beautifully gold, green, and blue colored when live, sometimes referred to as dolphin fish or *Dorado*. Found in tropical waters, its meat is mild and firm. Market size ranges from 15 to 20 pounds.

Marinade—A mixture used before cooking to flavor and moisten foods; may be liquid or dry. Liquid marinades are usually based on an acidic ingredient, such as wine or vinegar; dry marinades are usually salt based.

Midwater trawling—A fishing method in which a cone shaped net is submerged from the stern of a boat, and pulled through the water until full.

Mise en place—A French term referring to a general sense of organization and preparedness, both mental and physical, before starting a task in the kitchen. For fish fabrication it involves having the proper tools close at hand, water for cleaning the animals, and sharp knives.

Mollusk—One of a family of invertebrates consisting of seven classes including gastropods, bivalves, and cephalopods. Bodies of mollusks are generally unsegmented and often are protected by shells though physical characteristics vary by class and species.

Monkfish—A bottom-dweller with a flat, wide head and large pointed teeth, it is recognizable by an anterior dorsal fin-like projection that dangles from above its head to lure prey. Fished for on both sides of the Atlantic, it is in fact found worldwide. Its meat is firm with a mild shellfish flavor.

Mussel—Dark and shiny shelled bivalves, they are found worldwide as both wild and farm-raised. Their flesh is sweet and nutritious; only mussels that are closed firmly should be eaten and dead mussels should be discarded. They should be received in mesh sacks to prevent suffocation and prepared immediately.

Octopus—A cephalopod with eight tentacles, its flesh is firm and flavorful. Octopi are turned inside out during fabrication to remove the beak and guts. When cooked, its edible skin is easily peeled off for consumption.

Oyster—A variety of bivalve mollusk found in fresh or salt water. Filter-feeders, they are notably high in protein, vitamins, and essential nutrients, and are low in cholesterol. With a complexity of flavor comparable to wine, they should be soft and slippery, and are most flavorful when consumed raw.

Pan frying—A cooking method in which items are cooked in fat in a skillet. This generally involves more fat than sautéing or stir frying but less than deep frying.

Pathogens—Responsible for the development of food-borne illness, they are classified into four groups: bacteria, viruses, parasites, and fungi.

Paupiette—A fillet or scallopine of fish or meat that is rolled up, sometimes around a stuffing, and poached or braised.

Pellicle—A tacky skin which forms on a product allowed to air-dry after brining. This skin helps the smoke flavor and color adhere during the smoking process and also acts as a protective barrier.

Perch—A red-orange schooling fish found worldwide in both fresh and salt waters. Because they grow and mature slowly, they are currently in danger of being overfished. The fish are known for their mildly flavored, off-white fillets.

Purse seining—A fishing method where the animals are encircled by a net that is drawn closed and pulled up to the side of the boat where fish are scooped or pumped on board.

Quarter fillet—A method of flat fish fabrication in which fillets are removed from each side of the backbone on the top, and again on the bottom yielding four fillets in total.

Salmon—A firm, oily, mild flavored fish recognized by the pinkish hue of its skin. There are many species readily available in the Pacific and certain species are commonly farm-raised in the Atlantic. The tightly flaked flesh of the fish can be prepared using nearly any culinary technique; the high oil content of the fish lends it to curing and smoking.

Sauté—To cook quickly in a small amount of fat in a pan on the range top.

Scaling—Step in fish fabrication where scales are removed from the animal. A tool such as the back of a knife, spoon, or a specifically designed implement is used to brush against the scales, snapping them off to reveal the skin underneath.

Scallop—A bivalve with notably sweet and delicate flesh. Active swimmers, scallops cannot keep their shells closed and lose moisture quickly. Scallops require minimal cooking and are often sold at a high price in the market.

Sea ranching—A controversial aquaculture technique where large schools of fish are collected and herded together to a confined space within the sea. There they are fattened through controlled feeding for months before harvesting.

Sea urchin—A delicacy around the world, identifiable by their spiny, pin cushion-like physical exterior. The most desirable for consumption is the roe, the texture and appearance of which vary between the female and male sexes. Roe may be eaten raw or emulsified into a sauce.

Shallow poach—To cook gently in a shallow pan of barely simmering liquid. The liquid is often reduced and used as the base of a sauce.

Shellfish tags—Contain information about the product that is useful in dealing with potential food poisoning. Tags are kept on the fish until the entire package is emptied and then must be dated and kept for 90 days.

Shrimp—An arthropod found in both cold and warm ocean waters worldwide; the flavor and size varies between species. With a shelf life of only a few days, shrimp should be received and stored at low temperatures. May be found in numerous market forms from whole to headless, fresh to frozen, peeled, deveined, cooked, breaded and many more. Generally sold in counts expressed as a number of shrimp per pound.

Shrimp farming—An aquaculture practice, generally developed in wetlands or intertidal zones. While farm-raised shrimp may be less expensive, there are notable differences in production levels, size, and flavor when compared to wild shrimp.

Skate—Flat bodied and found worldwide, they average two feet in length from wing to wing, and average 10 pounds. The edible wings of the fish are best cooked a few days after the catch; the meat is similar in flavor to scallops and is string-like and juicy.

Snapper—Found in warm waters ranging from North Carolina to Brazil, the meat is valued for its delicate but firm texture. Over 125 species are in the family. They are reef fishes that can also live in rocky bottoms and ocean ledges. Its market size averages two to four pounds.

Sole—Often confused for flounder, they are distinguishable by their rounded heads and mouths, oval bodies, and small eyes. In North American waters there are six species, the most desirable of which is the Dover sole, a sweet and buttery flavored fish highly sought after worldwide and are generally very expensive.

Squid—Found in oceans worldwide, they are high in protein, low in fat, nearly 75% edible and average three to six inches in length. Their flesh is tender and mildly sweet, although it will become tough from overcooking.

Standard breading procedure—The assembly-line procedure in which items are dredged in flour, dipped in beaten egg, then coated with the crumbs before being pan fried or deep fried.

Steaks—Fish processed in cross-section cuts of flesh, generally with the bone in. Boneless steaks of some varieties are also available.

Steel—A tool used to sharpen knife blades. Long and thin with a handle at the bottom, they are usually made from steel.

Stew—A cooking method nearly identical to braising but generally involving smaller pieces of meat and hence a shorter cooking time. Stewed items may be blanched, rather than seared, to give the finished product a pale color. Also, a dish prepared by using the stewed method.

Stir-frying—A cooking method similar to sautéing in which items are cooked over very high heat, using little fat. Usually this is done in a wok, and the food is kept moving constantly.

Stone—A hard block used for knife sharpening, made from synthetic or natural materials, it is lubricated with water or oil.

Surimi—An Asian fish product where lean fish is puréed into a paste then enhanced with flavoring, color, starch, and other preservatives. Manipulated into various fish-like shapes, it can also be used to bind actual fish into authentic shapes of higher quality products.

Superchilling—A method of keeping seafood at a temperature below 32°F without freezing the product; superchilling significantly extends the given product's shelf life.

Swordfish—Found worldwide, they are identified by their sword-like snout. An average market size ranges from 50 to 225 pounds. Its flesh is sweet and firm, ideal for grilling due to its high fat content.

Temperature danger zone—Between 41°F and 135°F, the range in which pathogens thrive. Food should never be left within this temperature range for any amount of time over four hours.

Tomalley—Lobster liver, which is olive green in color and used in sauces and other items.

Up and over technique—A method of fabricating hard boned round fish where the fillet is removed by bringing the knife up and over the hard rib bones, rather than straight through.

Trawling—A method of fishing in which boats pull nets through the water in order to make a catch. The size of nets used varies; bottom trawlers also have chains that drag through the sea floor stirring up ground life to be caught in the net.

Trout—Commonly farm-raised, a delicately textured fish harvested at about eight to sixteen ounces. In the wild they are found in clean, cold water, and often migrate from rivers into the ocean. Good for numerous cooking preparations, it is often available smoked.

Tuna—An extremely powerful and fast fish, there are many species available worldwide, each purchasable in cuts with unique flavors and textures. Fatty and very rich in flavor, it is often consumed raw as well as sold canned throughout the United States.

Turbot—One of the largest round flat fish in the world, its body is rounded in a diamond-like shape. Species are found in the Pacific as well as in European waters and the seas around Greenland. European turbot is especially desirable, and has extremely white, mild, firm flesh.

Trolling—Using lures or baited lines weighted and pulled from the stern of a boat to catch valuable fish. A beneficial method as lines can be individually rigged for targeting specific species as well as can be drawn in quickly, retrieving the catch still alive and thus preserving quality in the caught fish.

Whole fillet—A flat fish fabrication technique in which a cut is made starting on an outer edge and working from the tail to the head on the top, and again on the bottom, yielding two fillets in total.

BIBLIOGRAPHY

G. Borgstrom, ed. *Fish as Food vol. 2: Nutrition, Sanitation, and Utilization* Academic Press, 1962.

The Culinary Institute of America. *The Professional Chef 8th Edition*. Wiley and Sons, Inc, 2006.

The Culinary Institute of America. *Techniques of Healthy Cooking*. Wiley and Sons, Inc. 2008.

Diversified Business Communications. *Seafood Handbook*. Diversified Business Communications, 2005.

A. Escoffier. *The Escoffier Cookbook: A Guide to the Fine Art of Cookery*. Crown Publishers, 1969.

J. D. Gilbert and C. R. Williams. *National Audubon Society Field Guide to Fishes of North America*. A Chanticleer Press Edition, Alfred A. Knopf, 2002.

S. T. Herbst. *The New Food Lover's Companion*. Barron's Educational series, 2007.

P. Johnson. *Fish Forever*. Wiley and Sons Inc., 2007.

M. Kurlansky, *Cod: A Biography of the Fish That Changed the World*. Penguin, 1997.

A. J. McClane. *The Encyclopedia of Fish Cookery*. Holt Rinehart and Winston, 1977.

N. Meinkoth. *The Audubon Society Field Guide to North America Seashore Creatures*. A Chanticleer Press Edition, Alfred A. Knopf, 1981.

P. Montagne. *Larousse Gastronomique: The Encyclopedia of Food, Wine and Cookery*. Crown Publishers Inc., 1961.

K. Omae and Y. Tachibana. *The Book of Sushi*. Kodansha International, Ltd., 1981.

C. R. Robins, G. C. Ray, G. Douglass. *Petersons Field Guides: A Field Guide to Atlantic Coast Fishes*. Houghton Mifflin Company, 1986.

R. Stein. *Rick Stein's Complete Seafood*. Ten Speed Press, 2004.

K. Whiteman. *The World Encyclopedia of Fish and Shellfish*. Lorenze Books, 2001.

C. A. Wright. *A Mediterranean Feast*. William Morrow and Company, 1999.

Montery Bay Aquarium
www.mbayaq.org
www.fishbase.org
United States Department of Agriculture
www.usda.gov
National Oceanic and Atmospheric Administration
www.noaa.gov
United States Department of Agriculture National Agricultural Library
www.nal.usda.gov
www.cfsan.fda.gov

INDEX

A

Abalone (*Haliotidae* family), 27–28, 117–118, 332
   as gastropods, 116
   market forms of, 28
   species of, 117
   storing, 28
   tenderizing, 117
Acid levels, 224
Aerobic bacteria, 208
Ahi. See Tuna: bigeye
Akami, 106, 241. *See also* Tuna
Alaskan king crab. *See* Crab: king
Algaculture, 195, 332
Amnesic shellfish poisoning, 214
Anaerobic bacteria, 208
Anchovies (*Engraulis encrasicolus, E. mordax*), 33–34, 332
   raw, 239
Anisakis, 208, 212
Ankimo, 70
Aquaculture, 193–197, 332. *See also* Farm-raised products
   shrimp farming, 195–197
   techniques in, 194–195
Aquaponics, 38–39, 194, 332. *See also* Farm-raised products
Arctic char. *See* Char, arctic
Arthropods, 115, 332
As-purchased (AP) cost, 15

B

Bacalao. See Cusk (*Brosme brosme*)
Bacteria, 208
Barnacles, 115
Barramundi (*Lates calcarifer*), 36
Bass, 332
   black sea (*Centropristis striata*), 41–42
   channel (*See* Drum, red (*Sciaenops ocellatus*))
   Chilean sea, 37–38
   European sea, 40–41
   hybrid, 38–40
   striped, 43–44
Bean curd, 243
Beeliner. *See* Snapper: vermilion
Belly burn, 12, 13
Birdseye, Clarence, 200, 203
Bisque, shrimp, 295
Bivalves, 116, 332
Blackfish (*Tautoga onitis*), 44–45
Bluefish (*Pomatomus saltatrix*), 45–46
Botulism, 207, 212
Bouillabaisse, 299–300
   monkfish in, 70
   mullet in, 72
Braising, 332
Branzino. See Bass, European sea
Breading, 336
Bream (*Sparus auratus*), 46–47
Brining, 218, 223, 332
   formula for, 227
   possible problems with, 227
Broiling, 332
Butterflying, 20, 183, 185, 332

C

Cape shark. *See* Dogfish (*Squalus acanthias*)
Carpaccio, 239
Carp roe, 237
Catfish, Atlantic. *See* Wolffish (*Anarchichas lupus*)
Catfish (*Ictalurus punctatus*), 47–48, 332
   farm-raising, 195
   sliminess of, 12
Caviar, 234–239. *See also* Roe
   grading, 235–236
   other than sturgeon, 236–237
   pressed, 236
   processing, 235
   quality characteristics of, 238
   serving, 238–239
   storing, 238
   varieties of, 235
Cephalopods, 116, 151–157, 332. *See also individual species*
   for sushi, 244
Certification, shellfish, 22–23
Ceviche, 224, 239, 249
   recipe for, 275
Channel bass. *See* Drum, red (*Sciaenops ocellatus*)
Char, arctic (*Salvelinus alpinus*), 35–36
Chilean sea bass. *See* Bass, Chilean sea
Chirashi sushi, 248
Chloramphenicol, 147
Chum salmon. *See* Salmon: chum
Ciguatera, 212–213

Clams, 24–25, 118–124, 332
    as bivalves, 116
    cherrystones, 120
    chowder, 120
    chowder, New England, 296
    counts of, 24
    fishing methods for, 9
    geoduck, 119–120
    hard-shell, 120–122
    IQF, 23
    knives for, 186
    littlenecks, 120
    market forms of, 24–25, 120
    opening, 180–181
    preparation of, 121
    purchasing, 22
    purging, 123
    quality characteristics of, 120–121, 122, 123, 124
    razor, 122
    removing dead, 24
    soft-shell, 122–123
    storing, 24
    surf, 124
    topnecks, 120
Clam sauce, 283
Clostridium botulinum, 212
Cod (Gadidae family), 49–56, 332
    Atlantic, 49–50
    black (*See* Sablefish *(Anoplopoma fimbria)*)
    buffalo, 66–67
    caviar, 236
    cheeks, 51
    curing, 221, 222
    cusk, 51–52
    fish and chips, 323
    fishing methods for, 4, 6, 7
    green, 66–67
    haddock, 52–53
    hake, 53–54
    Pacific, 55–56
    pan-fried, 321
    pollock, 54–55
    salt, 217, 222
    salt, fritters, 276
    scrod, 50–51
    whiting, 54
Coho salmon. *See* Salmon: Coho
Commission for the Conservation of Antarctic Marine Living Resources, 37–38

Conch, queen *(Strombus gigas)*, 28, 125–126, 332
    fritters, 277
    as gastropods, 116
    market forms of, 28
    shelf life of, 22
    tenderizing, 125
Cooling procedures, 210–211
Copper River salmon. *See* Salmon: sockeye
Coquilles St. Jacques, 178, 332
Costs
    as-purchased, 15
    edible portion, 15
    of fillets, 15
    shellfish fabrication, 173
Crab, 30–31, 126–132, 332–333
    blue, 126–128, 179
    classification of, 115
    crab cakes, 279–280
    determining quality of, 31
    Dungeness, 31, 129–130
    fabrication of, 179–180
    fishing methods for, 8
    Jonah, 31, 130–131
    king, 31, 128–129, 179
    market forms of, 31
    roe, 236
    shelf life of, 200
    snow, 31, 131
    soft-shell, 126, 128
    stone, 31, 131–132
    storing, 30
    in surimi, 149
Creole honey-mustard sauce, 281
Creole sauce, 125
Cross-contamination, 209–210, 333
Crustaceans, 333
Curing, 221–227, 333
    brining, 221, 223
    caviar, 234–239
    comparison of methods in, 225–226
    dry, 221, 222–223, 225–227, 333
    nitrates and nitrites, 224–225
    possible problems with, 227
    salt in, 223–224
    spices and herbs in, 225
    sweeteners in, 225
    wet, 225–226
Cusk *(Brosme brosme)*, 51–52
Cutting boards, 209
Cuttlefish *(Sepia officinalis)*, 116, 132–133

D

Dashi, 110–111, 294, 333
Daurade. See Bream *(Sparus auratus)*
Dealers, finding reputable, 205
Deep frying, 333
Deep poaching, 333
Dehydration, 223, 224
Denaturing of proteins, 223, 224
Dextrose, 225
Diarrheal shellfish poisoning, 214
Divers, shellfish collection by, 9
Docosahexaenoic acid (DHA), 195
Dogfish *(Squalus acanthias)*, 56–58
Dolphin fish. *See* Mahi mahi *(Coryphaena hippurus)*
Dolphins, fishing methods and, 5, 8, 105
Dorade. See Porgy *(Pagrus pagrus)*
Dorado. *See* Mahi mahi *(Coryphaena hippurus)*
Dover sole. *See* Sole *(Achiridae* family)
Dredging, 9, 333
Dried seafood, 217
   storage time for, 204
Drift nets, 5, 333
Drum, red *(Sciaenops ocellatus)*, 80, 81
Dublin bay prawns. *See* Langostino *(Cervimunida johni, Munida gregaria, Pleuroncodes monodon, Nephrops norvegius)*

E

E. coli, 207, 211–212
Echinoderms, 117
Edible portion (EP) cost, 14
Eel *(Anguilla rostrata)*, 58–59, 333
Egg pasta recipe, 284
Environmental issues
   abalone and, 27
   Chilean sea bass and, 37–38
   dolphins, 5, 8, 105
   factory ships, 200
   fishing methods and, 3, 4, 5, 105
   PCB contamination, 43–44
   sturgeon and, 234
   swordfish and, 98
Escabeche, 249
   hake, 53
Escoffier, Auguste, 96

F

Fabrication
   clam, 180–181
   crab, 179–180
   Dover sole, 170–171
   fin fish, 159–171
   flat fish, 161–164, 170–171
   lobster, 174–177
   mussel, 182
   oyster, 180–181
   round fish, 165–170
   scallop, 178
   sea urchin, 182
   shrimp, 183–186
   skate wing, 183
   squid, 177–178
   up and over technique, 336
Factory ships, 200
Facultative bacteria, 208
Farm-raised products
   abalone, 27–28
   barramundi, 37
   bass, hybrid, 38–39
   catfish, 47–48
   mussels, 25, 138
   oysters, 152
   salmon, 83, 84
   shrimp, 144, 147, 195–197
   tilapia, 100–101
   trout, 103
Fermentation, 223, 224, 241
FIFO, 205, 333
Fillet knives, 187
   scaling with, 161
Fillets, 333
   fabrication of flat fish, 161–164
   fabrication of round fish, 165–170
   freshness indicators for, 15
   poaching, 315–316
   portion-size, 16
   purchasing, 14–15
   quarter, 335
   skin in species identification of, 14, 15
   turbot, identifying, 111
   whole, 336
Fin fish. *See also individual species*
   fabrication of, 159–171
   flat, fabrication of, 161–164
   gutting, 159–160
   identification of, 33–113
   indicators of poor quality, 13
   market forms for, 14–16
   purchasing fresh, 11–13
   round, fabrication of, 161
   scaling, 160–161
   shelf life of, 200
   smell of fresh, 11–12, 13

storing fresh, 13
for sushi, 244
whole, 14–15
Finnan Haddie, 52
Fish. *See* Fin fish
Fish and chips, 323–324
Fishing methods, 3–9
  divers, 9
  dredging, 9
  drift nets, 5
  fish traps/pots, 8
  gillnetting, 4
  longlining, 4
  purse seining, 8
  trawling, 5–6, 7
  trolling, 6–7
Fish sauce, 33
Fish stock, 330
Fish traps/pots, 8, 92, 333
Flounder, 59–60, 333
  American plaice, 59
  fillets, 14–15
  fishing methods for, 6
  with white wine sauce, 316–317
  witch, 59
  yellowtail, 59
Flow-of-goods chain, 11
Flying fish roe, 236
Food and Drug Administration (FDA), 47, 56, 203, 206, 221
Food-borne diseases, 207–214, 333
  anisakis, 208, 212
  ciguatera, 212–213
  Clostridium botulinum, 212
  cross-contamination and, 209–210
  E. coli, 207, 211–212
  gempylotoxin, 212
  listeria, 212
  in oysters, 26
  pathogens in, 207–209
  red tide, 119, 213
  salmonella, 211
  scombroid, 213
  shellfish poisoning, 213–214
  symptoms of, 211
  tapeworms, 212
  temperature danger zone for, 210–211
Food safety
  cod and, 49
  HACCP and, 205–207
  lobster and, 136
  oysters, 153
  red tide and, 119
  sanitation and, 205–214
  shrimp farming and, 147
  storage and, 199–205
  sushi, 203, 243–244
Forcemeat, mousseline, 331
Freezing, 199–200. *See also* IQF (individually quick-frozen) products
  blast, 202
  cryogenic, 202–203
  ice, 201–202
  storage times for, 204
  for sushi, 244
  thawing and, 203, 204
French fried potatoes, 325
Freshness, 200–201
Fritters
  conch, 277
  salt cod, 276
Fungi, 207

G

Gastropods, 116, 333
Gempylotoxin, 212
General Foods Corporation, 200
Geoducks. *See* Clams: geoduck
Ghost nets, 5
Gillnetting, 4, 92, 333
Gills, color of on fresh fish, 12
Ginger, 243
  grilled salmon with ginger glaze, 318–320
Gloves, food-handling, 209–210
Government regulations, 206–207
Gravalox, 222, 226
  recipe for, 328
Grilling, 333
Grouper, red *(Epinephelus morio)*, 61–62
Gutting, 221, 333
  fin fish, 159–160

H

HACCP. *See* Hazard Analysis Critical Control Point (HACCP)
Hackleback caviar, 237
Haddock *(Melanogrammus aeglefinu)*, 52–53
Hake *(Merluccius merluccius)*, 53–54
Halibut *(Hippoglossus hippoglossus, H. stenolepis)*, 62, 333
  fillets, 14–15
  fishing methods for, 4

Hand washing, 209
Hawaiian moon fish. *See* Opah *(Lampris guttatus)*
Hazard Analysis Critical Control Point (HACCP), 201, 205–207, 334
   in preservation techniques, 218
Herring *(Clupea harengus)*, 63–64. *See also* Sardines *(Sardinella aurita, Sardina pilchardus, Harengula jaguana)*
   curing, 221
   raw, 239
Histamine poisoning, 213
Honey-mustard sauce, creole, 281
Honyaki knives, 188
Hygiene, 209–210. *See also* Sanitation

I

Ice, 199, 201–202, 334
   block, 201
   crushed, 202
   flake, 202
Infection, 334
International Commission for the North West Atlantic Fisheries, 200
In the round, 334
Intoxication, 334
IQF (individually quick-frozen) products
   clams, 25, 120
   conch, 28
   lobster, 30
   mollusks, 23
   mussels, 25
   oysters, 26
   pasteurization of, 26
   scallops, 27
   shrimp, 21

J

Japanese omelet, 243, 290
John Dory *(Zenopsis ocellata, Zeus faber)*, 65–66

K

Kasumi knives, 188
Katsu dashi, 110–111, 294, 333
Kazunoko (herring roe), 63
Knives, 186–190
   carbon steel, 186–187
   clam, 186
   fillet, 161, 187
   Japanese style, 188
   oyster, 186
   sashimi, 187
   sharpening, 187, 188–190
   stones for, 187, 336
   for sushi, 243

L

Langostino *(Cervimunida johni, Munida gregaria, Pleuroncodes monodon, Nephrops norvegius)*, 133–134, 200
Legal Seafood, 206–207
Lingcod *(Ophiodon elongatus)*, 66–67
Listeria, 212
Lobster, 29–30, 134–138, 334
   American, 134–137
   classification of, 115
   determining doneness of cooked, 136
   fabrication of, 174–177
   fishing methods for, 8
   killing, 177
   market forms of, 30
   purchasing, 29
   sizes and terminology of, 29–30
   soft-shell, 136
   spiny, 30, 137–138
   storing, 29
   in surimi, 149
   thermidor, 326
Lobsterette. *See* Langostino *(Cervimunida johni, Munida gregaria, Pleuroncodes monodon, Nephrops norvegius)*
Longline fishing, 4, 73, 334
Loup de mer. *See* Bass, European sea
Lumpfish roe, 237

M

Mackerel *(Scomber scombrus, Scomberomorous maculatur)*, 67–69, 334
   curing, 221
   fishing methods for, 6, 8
Mahi mahi *(Coryphaena hippurus)*, 69–70, 334
   fishing methods for, 4
Maki sushi, 247–248, 291–292
Mangrove snapper. *See* Snapper: gray
Marinade, 334
Market forms, 14–16
   fillets, 14–15
   steaks, 15–16
   whole fish, 14
Meunière, sole, 96, 306–308
Midwater trawling, 6, 7
Mirin, 243
Mise en place, 161, 222, 334

Miso soup, 70, 293
Mollusks, 22–29, 334. *See also* Abalone; Clams; Conch; Mussels; Oysters; Scallops
    breaded, 23
    certified, 22–23
    frozen, 23
    identification of, 115–116
    purchasing, 22–23
Monkfish *(Lophius americanus)*, 70–71, 334
    fishing methods for, 4
Mother of pearl, 117
Mousseline forcemeat, 331
Mullet, red *(Mugil cephalus)*, 72–73
    caviar, 236
Mussels, 334
    as bivalves, 116
    blue, 138
    cleaning, 25
    fabriction of, 182
    fishing methods for, 9
    green lip, 139
    IQF, 23
    marinière, 297
    market forms of, 25
    purchasing, 22
    storing, 25
    wild vs. farm-raised, 25
Mutton fish. *See* Snapper: mutton

N

National Marine Fisheries Service, 105
National Oceanic Atmospheric Administration (NOAA), 38
National Shellfish Sanitation Program, 206
Nematodes, in cod, 49
Neurologic shellfish poisoning, 214
Nigiri sushi, 245–247
    recipes for, 289–290
Nitrates/nitrites, 224–225
Nori-roll sushi recipe, 291–292
Nutritional information
    abalone, 118
    anchovies, 34
    arctic char, 36
    bass, black sea, 42
    bass, Chilean sea, 38
    bass, European sea, 41
    bass, hybrid, 40
    bass, striped, 44
    bluefish, 46
    bream, 47
    catfish, 48
    clams, geoduck, 120
    clams, hard-shell, 122
    clams, razor, 122
    clams, soft-shell, 123
    clams, surf, 124
    cod, Atlantic, 51
    cod, Pacific, 56
    conch, 126
    crab, blue, 126, 128
    crab, Dungeness, 130
    crab, Jonah, 131
    crab, king, 129
    crab, snow, 131
    crab, soft-shell, 128
    crab, stone, 132
    cusk, 52
    cuttlefish, 133
    dogfish, 58
    eel, 59
    grouper, red, 62
    haddock, 53
    hake, 54
    halibut, 62
    herring, 64
    John Dory, 66
    king, 128–129
    langostino, 134
    lingcod, 66
    lobster, Maine, 137
    lobster, spiny, 138
    mackerel, 69
    mahi mahi, 70
    mullet, 73
    mussels, blue, 138
    mussels, green lip, 139
    ocean perch, 76
    octopus, 152
    opah, 74
    orange roughy, 76
    oysters, Eastern, 154
    oysters, European, 155
    oysters, Pacific, 156
    pollock, 55
    pompano, 78
    porgy, 80
    sablefish, 82
    salmon, Atlantic, 84
    salmon, chum, 86
    salmon, Coho, 84

Nutritional information *(Continued)*
   salmon, sockeye, 86
   salt, 220–221
   scallops, bay, 141
   scallops, sea, 142
   sea urchin, 144
   shrimp, Gulf, 146–147
   shrimp, Pacific white, 148
   shrimp, pink, 148
   shrimp, rock, 149
   shrimp, tiger, 145
   skate, 90
   smelt, 92
   snapper, red, 93
   sole, 97
   squid, 157
   sturgeon, 98
   surimi, 149
   swordfish, 100
   tilapia, 100
   tilefish, 102
   trout, 104
   tuna, albacore, 106
   tuna, bigeye, 108
   tuna, bluefin, 109
   tuna, bonito, 110
   tuna, yellowfin, 110
   turbot, 112
   wolffish, 113

O

Octopus *(Octopus dofleini, O. vulgaris)*, 116, 151–152, 334
Omega-3 fatty acids, 33, 195
Opah *(Lampris guttatus)*, 73–74
Orange roughy. *See* Roughy, orange *(Hoplostethus atlanticus)*
Osmosis, 223, 224
Oursins. *See* Sea urchins *(Strongylocentrotus fransiscanus, S. droebachiensis)*
Overhauling, 221
Oysters, 25–26, 152–156, 334
   as bivalves, 116
   counts, 26
   Eastern, 154
   European, 154–155
   fishing methods for, 9
   food safety and, 153
   IQF, 21
   knives for, 186
   kumomoto, 155
   market forms of, 26
   olympia, 155
   opening, 180–181
   Pacific, 155–156
   pasteurized, 26
   purchasing, 22, 25
   storing, 25–26

P

Paella, 310–313
Pan frying, 334
Paralytic shellfish poisoning, 213
Parasites, 208, 212
Pasteurization
   caviar, 238
   crab meat, 130
   oyster, 26
Patagonian tooth fish. *See* Bass, Chilean sea
Pathogens, 334
Paupiette, 335
PCBs (polychlorinated biphenyl), 43–44
Pellicle, 218, 228, 335
Perch, Atlantic ocean *(Sebastes marinus)*, 76–77, 335
Periwinkles, 116
Permit *(Trachinotus falcatus)*, 77
Pillsbury Corporation, 205
Plaice, American *(Hippoglossoides platessoides)*, 59
Pliers, 186
Poached fish recipe, 314–316
Poaching, 333, 335
Poke, 239
Pollock *(Pollachius virens)*, 54–55
   fishing methods for, 6, 7
Pompano *(Trachinotus carolinus)*, 77–78
Ponds, aquaculture, 195
Porgy *(Pagrus pagrus)*, 79–80
Portion size
   fillets, 15
   steaks, 16
Potentially hazardous foods, 208–209
Prawns. *See* Langostino *(Cervimunida johni, Munida gregaria, Pleuroncodes monodon, Nephrops norvegius)*
Preservation, 217–249. *See also* Storage
   salt in, 217, 218–227
   smoking, 228–233
Protein denaturing, 223, 224

Purchasing, 205
    fresh fin fish, 11–13
    quality characteristics for, 11–13
Purging solution, 123
Purse seining, 8, 105, 335
Pyloric caeca, 221

Q

Quarter fillets, 335

R

Raceways, 194–195
Raw seafood, 239–241
    sashimi, 106, 120, 239–241
    sushi, 108, 203, 218, 241–249
Razor clams. *See* Clams: razor
Recipes, 273–330
    brine for seafood, 227
    ceviche, 274–275
    clam purging solution, 123
    clam sauce, 283
    cold-smoked salmon, 327
    conch fritters, 277
    crab cakes, 278–280
    creole honey-mustard sauce, 281
    Dover sole meunière, 306–308
    dry cure for salmon, 226
    egg pasta, 284
    fish and chips, 322–324
    fish stock, 330
    flounder with white wine sauce, 316–317
    french fried potatoes, 325
    garlic shrimp, 282
    gravlax, 226, 328
    grilled salmon with ginger glaze, 318–320
    Japanese hand vinegar, 288
    lobster thermidor, 326
    miso soup, 293
    mousseline forcemeat, 331
    mussels marinière, 297
    Nigiri sushi, 289
    paella, 310–313
    pan-fried cod, 321
    rouille, 301
    salade niçoise, 302–304
    salmon rillette, 329
    salt cod fritters, 276
    shallow poached fish, 314–316
    shrimp bisque, 295
    shrimp tempura, 285–286
    sushi rice, 241–242, 287–288
    tempura dipping sauce, 286
    trout with sautéed mushrooms, 305
    wasabi, 288
Recirculation systems, 194
Record keeping
    for certified shellfish, 22–23
    for smoked seafood, 233
Red drum. *See* Drum, red *(Sciaenops ocellatus)*
Redfish. *See* Drum, red *(Sciaenops ocellatus)*
Red fish. *See* Perch, Atlantic ocean *(Sebastes marinus)*
Red tide, 119, 213
Refrigeration, 205
    temperatures for, 211
Reheating, 211
Repertoire de la Cuisine, Le (Escoffier), 96
Rice, for sushi, 241–242, 287–288
Rice paper sheets, 242
Roe
    caviar, 234–239
    herring, 63
    kazunoko, 63
    lobster, 136
    mullet, 72
    salmon, 85
    sea urchin, 28–29, 143, 182
    sturgeon, 98
Rollmops, 63
Rouget. *See* Mullet, red *(Mugil cephalus)*
Roughy, orange *(Hoplostethus atlanticus),* 74–76
Rouille, 301

S

Sablefish *(Anoplopoma fimbria),* 81–82
Salade niçoise, 302–304
Salmon, 335
    Atlantic, 82–84
    caviar, 236, 237
    chum, 85–86, 87
    Coho, 84
    cold-smoked, 327
    curing, 222, 223, 226, 227
    fillets, 14
    fishing methods for, 5
    gravlax, 222, 226, 328
    grilled, with ginger glaze, 318–320
    pink, 86, 87
    raw, 239
    rillette, 329
    smoked, curing, 223, 327

Salmon *(Continued)*
   smoking, 231
   sockeye, 86, 87
   for sushi, 244
Salmon Control Plan, 206
Salmonella, 211
Salmonellosis, 207
Salt, 217, 218–227
   in brines, 223
   cooking with, 219–220
   curing with, 221–227
   health and, 220–221
   history of, 220
   kosher, 219–220, 223
   manufacture of, 218
   sea, 219, 220
   types of, 220
Saltpeter, 224–225
Sanitation, 205–214
   cross-contamination and, 209–210
   food-borne disease and, 207–214
   HACCP and, 205–207
Sardines *(Sardinella aurita, Sardina pilchardus, Harengula jaguana)*, 63, 88–89
Sashimi
   geoduck, 120
   tuna for, 106, 239–241
Sautéing, 335
Scalers, 186
Scales
   quality of on fresh fish, 12
   removing, 160–161, 186, 335
Scallops, 26–27, 139–142, 335
   bay, 140–141
   as bivalves, 116
   counts, 27
   dry curing, 227
   fabrication of, 178
   fishing methods for, 9
   IQF, 23
   market forms of, 27
   purchasing, 22, 26–27
   sea, 141–142
   wet vs. dry, 27, 140, 142
Scampi. *See also* Langostino *(Cervimunida johni, Munida gregaria, Pleuroncodes monodon, Nephrops norvegius)*
Scombroid, 213
Scrod, 50–51
Sea bream. *See* Porgy *(Pagrus pagrus)*

Sea cucumbers, 117
Sea eggs. *See* Sea urchins *(Strongylocentrotus fransiscanus, S. droebachiensis)*
Sea ranching, 194, 335
Sea urchins *(Strongylocentrotus fransiscanus, S. droebachiensis)*, 28–29, 117, 142, 335
   fabriction of, 182
   market forms of, 29
   roe from, 28–29, 143, 236
   sauce from, 143
Seaweed, nori, 242
Shark, cape. *See* Dogfish *(Squalus acanthias)*
Shelf life, 200–201, 204
Shellfish, 19–31. *See also individual species*
   arthropods, 115
   bivalves, 116
   cephalopods, 116
   certified, 22–23
   classification of, 115–117
   demand for, 19
   echinoderms, 117
   fabrication of, 173–186
   fishing methods for, 9
   gastropods, 116
   identification of, 115–149
   storing, 23
   for sushi, 244
Shellfish poisoning, 213–214
Shellfish tags, 22–23, 26, 335
Shrimp, 144–149, 335
   bisque, 295
   black tiger, 145, 195, 196
   blocks of, 21
   breaded, 21
   butterflied, 20
   butterflying, 183, 185
   classification of, 115
   cooked market forms, 21
   counts and sizes of, 21–22
   dry curing, 227
   fabrication of, 183–186
   farm-raising, 195–197, 335
   fishing methods for, 6
   fresh, 144–145
   frozen, 21
   garlic, 282
   green, headless, 20
   Gulf, 146–147
   market forms of, 20–22
   in nigiri sushi, 246–247

Pacific white, 147–148
peeled, 20
peeled and deveined, 20
peeling and deveining, 183, 184
pieces, 21
pink, 148
purchasing, 19
recipes, 285–286, 295, 309–313
rock, 148–149
shelf life of, 200
shell-on and cooked, 20
storing, 20
in surimi, 149
for sushi, 246–247
tempura, 186, 285–286
with tomatoes, feta, and oregano, 309
whole, head-on, 20
wild vs. farm-raised, 144, 197
Skate *(Raja batis, R. binoculata, Gymnura micrura)*, 89–90, 335
fabrication of, 183
sliminess of, 12
Slime fish. *See* Roughy, orange *(Hoplostethus atlanticus)*
Smelt, rainbow *(Osmerus mordax)*, 91–92
Smoked seafood, 218
benefits of, 230
cold-smoked salmon, 327
cold smoking, 230–231, 233
hot smoking, 231–233
pellicle formation on, 228
preparing, 228–233
storage time for, 204
processing analysis for preparing, 233
woods for, 228–229
Snails, 116
Snapper *(Lutjanidae* genus), 92–95, 335
gray, 93
mutton, 95
ocean perch compared with, 76
red, 14, 76, 92–93, 94
vermilion, 94, 95
yellowtail, 94, 95
Sockeye salmon. *See* Salmon: sockeye
Sodium tripolyphosphate (STP), 27, 140, 142
Sole *(Achiridae* family), 95–97, 335
Dover, 95–96, 170–171, 306–308
fabrication of, 170–171
fishing methods for, 6
lemon, 95

meunière, 96, 306–308
petrale, 95
Soups and stocks
bouillabaisse, 298–300
clam chowder, New England, 296
dashi, 110–111, 294, 333
fish, recipe for, 330
lobster shells in, 174
miso, 293
shrimp bisque, 295
shrimp shells in, 183
storage times for, 204
Soy sauce, 243
Squid *(Loligo illecebrosus, L. opalescens, L. pealei)*, 156–157, 336
as cephalopod, 116
fabrication of, 177–178
fishing methods for, 5
ink, 156
collecting, 177
market forms of, 31
purchasing, 31
sliminess of, 12
St. Peter's fish. *See* John Dory *(Zenopsis ocellata, Zeus faber);* Tilapia *(Tilapia nilotica)*
Steaks, 15–16, 336
preparing, 16
Steels, 336
Stir-frying, 336
Stocks. *See* Soups and stocks
Stones, knife sharpening, 187, 336
Storage, 199–205, 204–205
cross-contamination and, 209
FIFO in, 205
food safety and, 211
temperature fluctuations during, 200, 201
Stripers. *See* Bass, striped *(Morone saxatilis)*
Sturgeon *(Acipenser medirostris, A. transmontanus)*, 97–98
caviar, 234–239
types of, 234
Super-chilling, 201, 336
Surimi, 21, 149, 336
Sushi, 241–249
chirashi, 248
frozen, 203
Japanese omelet, 243
key ingredients in, 243–244
maki, 247–248, 291–292
Nigiri, 245–247, 289–290

Sushi *(Continued)*
   nori-roll, 291–292
   rice for, 241–242, 287–288
   safety of, 218
   temaki, 248
   tuna for, 108
   wrappings for, 242–243
Swordfish *(Siphias gladius)*, 98–100, 336
   fishing methods for, 4

T

Tags, shellfish certification, 22–23, 26, 335
Taniguchi method, 240
Tapeworms, 212
Tautogs. *See* Blackfish
Temaki, 248–249
Temperature danger zone, 210–211, 336
Tempura, shrimp, 186, 285–286
Thawing frozen seafood, 204
Tilapia *(Tilapia nilotica)*, 100–101
   farm-raising, 195
Tilefish *(Lopholatilus chamaeleonticeps)*, 101–102
Tobiko (flying fish roe), 237
Tomalley, 136, 336
Toro, 105–106, 240–241. *See also* Tuna
Tranches, 167
Trawling, 5–6, 7, 92, 334, 336
Trichinella spiralis, 208
Trolling, 6–7, 336
Trout, rainbow *(Salmo gairdneri)*, 103–104, 336
   caviar, 237
   dry curing, 227
   farm-raising, 195
Trout with Sautèed Mushroom, 305
Tuna, 104–111, 336
   albacore, 106
   aquaculture of, 193
   Atlantic bluefin, 108
   bigeye, 107–108
   blackfin, 108
   bonito, 110–111
   caviar, 236
   curing, 221
   fishing methods for, 4, 5, 7, 8, 104, 105
   in katsu dashi, 110–111, 294, 333
   longtail, 108
   in nigiri sushi, 245–247
   overfishing of, 104
   overview of, 104–106
   in salade niçoise, 303–304
   for sashimi, 239–241
   Scombridae family, 106
   southern bluefin, 108–109
   for sushi, 203, 244, 245–247
   tartare, 239
   Thunnus family, 106
   yellowfin, 109–110
Turbot *(Scophthalmus maximus, Reinhardtius hippoglossoides)*, 111–112, 336
Tweezers, 186

U

Uniformity ratios, 21–22
Uni (sea urchin roe), 28–29, 143, 237. *See also* Sea urchins *(Strongylocentrotus fransiscanus, S. droebachiensis)*
Up and over technique, 336
U.S. Customs, 38
U.S. Department of Agriculture (USDA), 119, 205, 206, 240

V

Vacuum-pack bags, 203–204
Vana, 29
Vegetables, for sushi, 243
Vibrio vulnificus, 26
Vinegar, in sushi, 243, 288
Viruses, 207

W

Wasabi, 243, 288
Whelk, 116
Whitefish caviar, 236, 237
White wine sauce, fillet of flounder with, 316–317
Whiting *(Merlangius merlangus)*, 54
Witch flounder *(Glyptocephalus cynoglossus)*, 59
Wolffish *(Anarchichas lupus)*, 112–113
Wood, for smoking foods, 228–230
Worcestershire sauce, anchovies in, 33

Y

Yellowtail flounder *(Limanda ferruginea)*, 59

PHOTO CREDITS

KEITH FERRIS

Pages 12, 13, 14 (bottom), 15, 20, 24, 25, 26, 29 (top), 30, 81, 160, 161, 162, 163, 164, 165, 166, 167, 168, 169, 170, 171, 174, 175, 176, 177, 178, 179, 180, 181, 182, 183, 184, 185, 186, 189, 190, 197 (bottom), 202, 222, 223, 231, 235, 236, 237, 238, 240, 242, 245, 246, 247, 248, 249, 264

BEN FINK

14 (top), 21, 27, 28, 29 (bottom), 34, 35, 36, 37, 39, 41, 42, 43, 45, 47, 48, 50, 51, 52, 53, 54, 55, 57, 58, 60, 61, 62, 64, 65, 67, 68, 69, 71, 72, 75, 77, 78, 79, 82, 83, 85, 87, 88, 90, 91, 94, 96, 99, 101, 102, 103, 110, 112, 113, 118, 119, 121, 122, 123, 124, 125, 127, 128, 129, 130, 131, 132, 133, 135, 137, 139, 140, 141, 142, 143, 145, 146, 147, 148, 149, 152, 153, 154, 155, 157, 176, 187, 197 (top), 219, 230, 274, 278, 280, 298, 300, 302, 304, 306, 308, 310, 312, 313, 314, 316, 318, 320, 322, 324

CIA CONVERSION CHARTS

TEMPERATURE, WEIGHT AND VOLUME CONVERSIONS

TEMPERATURE CONVERSIONS

32°F = 0°C	205°F = 96°C	380°F = 193°C
35°F = 2°C	210°F = 99°C	385°F = 196°C
40°F = 4°C	**212°F = 100°C**	390°F = 199°C
45°F = 7°C	215°F = 102°C	395°F = 202°C
50°F = 10°C	220°F = 104°C	**400°F = 204°C**
55°F = 13°C	**225°F = 107°C**	405°F = 207°C
60°F = 16°C	230°F = 110°C	410°F = 210°C
65°F = 18°C	235°F = 113°C	415°F = 213°C
70°F = 21°C	240°F = 116°C	420°F = 216°C
75°F = 24°C	245°F = 118°C	**425°F = 218°C**
[room temp]	**250°F = 121°C**	430°F = 221°C
80°F = 27°C	255°F = 124°C	435°F = 224°C
85°F = 29°C	260°F = 127°C	440°F = 227°C
90°F = 32°C	265°F = 129°C	445°F = 229°C
95°F = 35°C	270°F = 132°C	**450°F = 232°C**
100°F = 38°C	**275°F = 135°C**	455°F = 235°C
105°F = 41°C	280°F = 138°C	460°F = 238°C
110°F = 43°C	285°F = 141°C	465°F = 241°C
115°F = 46°C	290°F = 144°C	470°F = 243°C
120°F = 49°C	295°F = 146°C	**475°F = 246°C**
125°F = 52°C	**300°F = 149°C**	480°F = 249°C
130°F = 54°C	305°F = 152°C	485°F = 252°C
135°F = 57°C	310°F = 154°C	490°F = 254°C
140°F = 60°C	315°F = 157°C	495°F = 257°C
145°F = 63°C	320°F = 160°C	**500°F = 260°C**
150°F = 66°C	**325°F = 163°C**	505°F = 263°C
155°F = 68°C	330°F = 166°C	510°F = 266°C
160°F = 71°C	335°F = 168°C	515°F = 268°C
165°F = 74°C	340°F = 171°C	520°F = 271°C
170°F = 77°C	345°F = 174°C	**525°F = 274°C**
175°F = 79°C	**350°F = 177°C**	530°F = 277°C
180°F = 82°C	355°F = 179°C	535°F = 279°C
185°F = 85°C	360°F = 182°C	540°F = 282°C
190°F = 88°C	365°F = 185°C	545°F = 285°C
195°F = 91°C	370°F = 188°C	550°F = 288°C
200°F = 93°C	**375°F = 191°C**	

WEIGHT CONVERSIONS

For weights less than 1/4 oz: use tsp/tbsp for U.S. measure with gram or mL equivalent (see specific conversion tables).

Formula to convert ounces to grams: number of oz × 28.35 = number of grams (round up for .50 and above)

1/4 ounce	7 grams
1/2 ounce	14 grams
1 ounce	28.35 grams
4 ounces	113 grams
8 ounces (1/2 pound)	227 grams
16 ounces (1 pound)	454 grams
32 ounces (2 pounds)	907 grams
40 ounces (2 1/2 pounds)	1.134 kilograms

VOLUME CONVERSIONS

Formula to convert fluid ounces to milliliters: number of fluid ounces × 30 = number of milliliters

1/2 fl oz	15 mL	20 fl oz	600 mL
1 fl oz	30 mL	24 fl oz	720 mL
1 1/2 fl oz	45 mL	30 fl oz	900 mL
1 3/4 fl oz	53 mL	**32 fl oz**	**960 mL [1 qt]**
2 fl oz	**60 mL**	40 fl oz	1.20 L
2 1/2 fl oz	75 mL	44 fl oz	1.32 L
3 fl oz	90 mL	**48 fl oz**	**1.44 L [1 1/2 qt]**
3 1/2 fl oz	105 mL	64 fl oz	1.92 L [2 qt]
4 fl oz	120 mL	**72 fl oz**	**2.16 L [2 1/2 qt]**
5 fl oz	150 mL	80 fl oz	2.4 L
6 fl oz	180 mL	96 fl oz	2.88 L [3 qt]
7 fl oz	210 mL	128 fl oz	3.84 L [1 gal]
8 fl oz	**240 mL [1 cup]**	1 1/8 gal	4.32 L
9 fl oz	270 mL	1 1/4 gal	4.8 L
10 fl oz	300 mL	1 1/2 gal	5.76 L
11 fl oz	330 mL	**2 gal**	**7.68 L [256 fl oz]**
12 fl oz	360 mL	3 gal	11.52 L
13 fl oz	390 mL	4 gal	15.36 L
14 fl oz	420 mL	5 gal	19.20 L
15 fl oz	450 mL	10 gal	38.40 L
16 fl oz	**480 mL [1 pt]**	20 gal	76.80 L
17 fl oz	510 mL	25 gal	96 L
18 fl oz	540 mL	50 gal	192 L
19 fl oz	570 mL